Make:
Tech DIY

Make: Tech DIY

Easy Electronics Projects *for* Parents *and* Kids

Ji Sun Lee and Jaymes Dec

Printed in Canada.

Published by Maker Media, Inc., 1160 Battery Street East, Suite 125, San Francisco, California 94111.

Maker Media books may be purchased for educational, business, or sales promotional use. Online editions are also available for most titles (safaribooksonline.com). For more information, contact our corporate/institutional sales department: 800-998-9938 or corporate@oreilly.com.

Publisher: Roger Stewart
Editor: Rebecca Rider, Happenstance Type-O-Rama
Proofreader: Elizabeth Welch, Happenstance Type-O-Rama
Interior Designer and Compositor: Maureen Forys, Happenstance Type-O-Rama
Cover Designer: Maureen Forys, Happenstance Type-O-Rama
Indexer: Valerie Perry, Happenstance Type-O-Rama

September 2016: First Edition

Revision History for the First Edition

2016-08-19 First Release

See oreilly.com/catalog/errata.csp?isbn=9781680451771 for release details.

978-1-6804-5177-1

Safari® Books Online

Safari Books Online is an on-demand digital library that delivers expert content in both book and video form from the world's leading authors in technology and business.

Technology professionals, software developers, web designers, and business and creative professionals use Safari Books Online as their primary resource for research, problem-solving, learning, and certification training.

Safari Books Online offers a range of plans and pricing for enterprise, government, education, and individuals. Members have access to thousands of books, training videos, and prepublication manuscripts in one fully searchable database from publishers like O'Reilly Media, Prentice Hall Professional, Addison-Wesley Professional, Microsoft Press, Sams, Que, Peachpit Press, Focal Press, Cisco Press, John Wiley & Sons, Syngress, Morgan Kaufmann, IBM Redbooks, Packt, Adobe Press, FT Press, Apress, Manning, New Riders, McGraw-Hill, Jones & Bartlett, Course Technology, and hundreds more. For more information about Safari Books Online, please visit us online.

How to Contact Us

Please address comments and questions concerning this book to the publisher:

Make:
1160 Battery Street East, Suite 125
San Francisco, CA 94111
877-306-6253 (in the United States or Canada)
707-639-1355 (international or local)

Make: unites, inspires, informs, and entertains a growing community of resourceful people who undertake amazing projects in their backyards, basements, and garages. Make: celebrates your right to tweak, hack, and bend any technology to your will. The Make: audience continues to be a growing culture and community that believes in bettering ourselves, our environment, our educational system—our entire world. This is much more than an audience; it's a worldwide movement that Make: is leading and we call it the Maker Movement.

For more information about Make:, visit us online:

- Make: magazine makezine.com/magazine
- Maker Faire makerfaire.com
- Makezine.com makezine.com
- Maker Shed makershed.com
- To comment or ask technical questions about this book, send email to bookquestions@oreilly.com.

The authors would like to dedicate this book to our good friend Robert Moon, who loved working with technology, crafts, and children.

Acknowledgments

Special thanks to Ji-Sun Lee's daughter Hannah Kim, without whom this book would not exist.

Thanks to our family, friends, and colleagues for their support: Chris Hyun-Chul Kim, Baek-Young Lee, Myung-Ja Na, Ji-Hyun Lee, Sang-Hoon Lee, Young-In Na, Kyung-Jin Jang, Cindy Seungwan Yoo, Gretchen Dec, Hsing Wei, Oya Kosebay, Lesa Wang, Maureen Reilly, Eric Walters, and Concepcion Alvar. Thank you to our publisher, Roger Stewart, our editor Rebecca Rider, and our book designer, Maureen Forys.

Contents

Preface

The idea for this book came during Ji Sun Lee's first visit to Maker Faire in 2007. Held at the San Mateo County Event Center in California, this was the second annual festival of making, craft, and technology projects organized by Make Media. Ji Sun was at the Faire to exhibit a project called the "Interactive Cake." Inspired by her then two-year-old daughter building cakes with her blocks, this project was a three-dimensional puzzle that formed a cake when assembled. Electronics embedded in the cake lit up, played songs, and even sensed when the candles were blown out.

While at the Maker Faire, Ji Sun noticed that many of the projects on display were not as appealing to girls as they were to boys. She also reflected on the fact that in her 15 years of working in information technology, she had met very few other women in the field. In response, she came up with the idea for *Tech DIY*, a series of sewing circuit projects that are meant to

attract and teach technology to girls and their mothers. Later on, Tech DIY expanded its audience to include all kids and adult electronics hobbyists. This book is the culmination of nine years of research and workshops with Tech DIY projects.

Why This Book?

There was a time when kids could take apart the technological world around them. In garages, basements, and other makeshift makerspaces, children would disassemble, poke around, and occasionally reassemble radios, telephones, and VCRs. Many influential scientists and engineers attribute their desire to enter their professions to tinkering around and playing with the parts of these devices to discover how they work.

Sadly, these opportunities to take apart and then repurpose, or hack, contemporary technologies are becoming less frequent. As modern devices shrink in size, they are made up of more embedded circuits and integrated electronics. Good luck finding any screws to open your modern smart phone, for instance. And if you do expose the circuits in your laptop, it is very difficult to isolate and reuse the tiny components; and besides, by doing so, you just voided your warranty. The fact is that most new technologies are not meant to be taken apart, or repurposed. They are designed to become obsolete and be discarded for the "next great thing."

At the same time, we live in a world where learning about technology is important for everyone, not just budding computer scientists or electronic engineers. We are surrounded by technology and it plays an increasingly important role in our lives. If children don't have opportunities to tinker and play with technology, they become more alienated and unfamiliar with how their world works. Thankfully, there is a movement afoot that could solve this dilemma.

Many people around the world are exploring creative applications of technology that often combine traditional crafts with electronics or computing. Supported by open source hardware and software, as well as communities and websites that specialize in sharing projects, the Maker Movement is helping people use technology to challenge, shape, and change the world around them. Technological materials and tools are increasingly becoming less expensive and more accessible to children.

The authors of this book are both educators who are passionate about empowering students with the abilities and skills needed to express themselves with technology. We hope that by introducing children to circuits through familiar crafts and materials like sewing and thread, we will allow them to feel more comfortable and confident when they approach and learn about electronics.

Who Is This Book For?

This book is for children, parents, and educators who want to learn about electronics and enjoy working with crafts and soft materials, like sewing and embroidery. We have found that children over the age of 10 are able to finish these projects with minimal help.

This book is not simply a series of craft projects. Our goal is to teach our readers about the basic principles of working with technology, from electronics to programming. You will learn how to turn on an LED, make a switch, understand series and parallel circuits and the relationship between electric current and voltage, learn how to use sensors and integrated circuits, and learn how to use solar energy. Since this book is aimed at children, we chose not to use solder to make the circuits in this book. Instead, the projects use conductive thread, eliminating the risk of respiratory problems or burns.

Chapter 1

Before You Start

Before beginning any project, you'll want to gather all of the necessary tools and materials. In this chapter, we introduce some of the more common supplies that you need when you're sewing circuits. We also describe some of the basic stitches that are used in embroidery and sewing.

Needlework Tools and Materials

Figure 1.1 shows the most frequently used tools and materials for sewing and embroidery. You can purchase these items at any hobby or craft store or online.

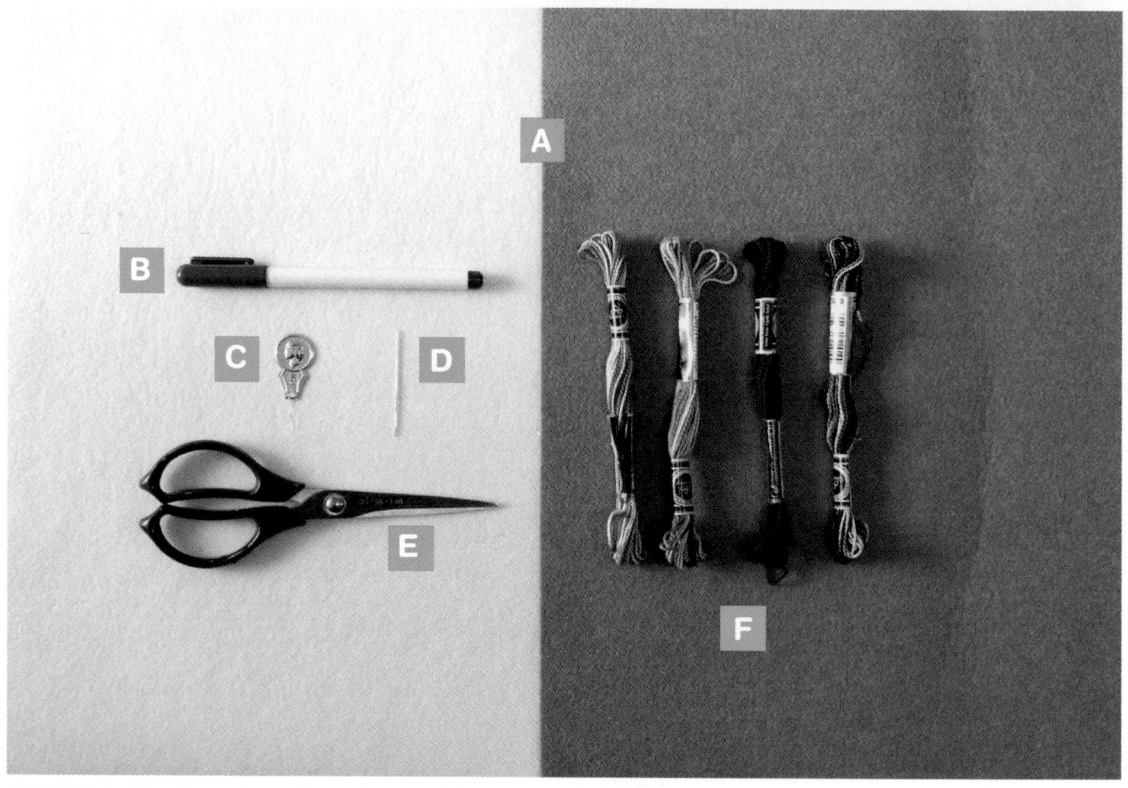

FIGURE 1.1: Needlework tools and materials

A **Felt**—It is easy to find felt in various colors and sizes. Look for it at any hobby shop or online craft supply store. If felt is too expensive, you can use any other fabric, including old clothing, for these projects.

B **Chalk pen**—Chalk pen markings are easily removable. You can purchase these pens anywhere that sells sewing supplies.

C **Needle threader**—It can be really difficult to thread a needle, especially with embroidery floss or conductive thread. Make sure you have several needle threaders available because they can get damaged easily.

D **Needle**—A needle with a larger eye is easier to thread. If a child will be using the needle, it is better to start off with a blunt embroidery needle rather than a regular sewing needle. However, when you are sewing with conductive thread, you will need to use a #5 needle or smaller to get it to go through the holes in the battery holder.

E **Scissors**—Most projects in this book use sheets of felt, so common paper scissors will work fine. However, fabric scissors work better on thinner fabric.

F **Embroidery floss**—Embroidery floss, or stranded cotton, is made up of six threads that are twisted together. It comes in hundreds of colors and you can buy it anywhere that sells craft, hobby, or sewing supplies.

Basic Needlework: Knots and Stitches

The following are a few stitches and knots you should know in order to use embroidery or sew circuits. Feel free to refer back to these instructions as you work on the projects in this book. Many great video tutorials online also demonstrate these stitches.

TYING THREAD AT THE BEGINNING

When you start sewing, tie a simple overhand knot at the end of your thread to keep it from passing through the fabric.

TYING THREAD AT THE END

When you are done with a line of stitches, poke the needle about halfway through a bit of the fabric. Wrap the thread around the sharp end of the needle two or three times (see Figure 1.2). Hold the thread and fabric still as you pull the needle through the loops. Make sure your knot is nice and tight before you cut off the loose end of the thread (see Figure 1.3). If you are sewing with conductive thread, it is a good idea to put a dab of clear nail polish or fabric glue on your knot to keep it from unraveling.

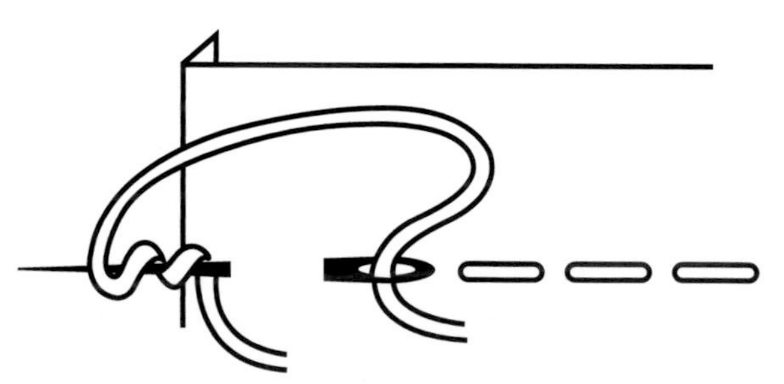

FIGURE 1.2: Tying the thread at the end

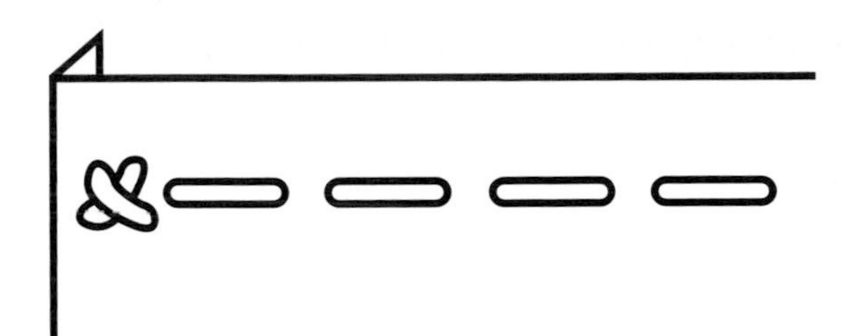

FIGURE 1.3: Finished knot at the end of your thread

THE RUNNING STITCH

This is the most basic of stitches. Poke the needle over and then under the fabric at regular intervals. You can make your stitches large or small depending on how you want the stitch to look.

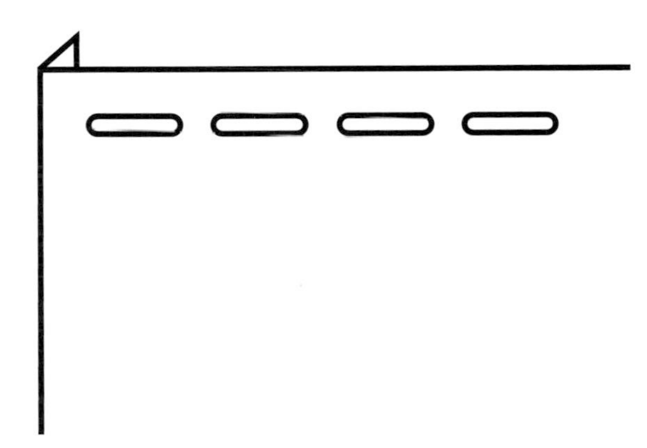

THE BACK STITCH

This is the most commonly used stitch in embroidery. It can look beautiful with colored thread. When you poke your nee-dle up to begin a stitch, move the needle back to poke it down through the hole where your thread went down on your last

stitch. After that, stitch out farther forward from the last time you poked your needle up, and continue that process of going back to the end of your last stitch before moving forward.

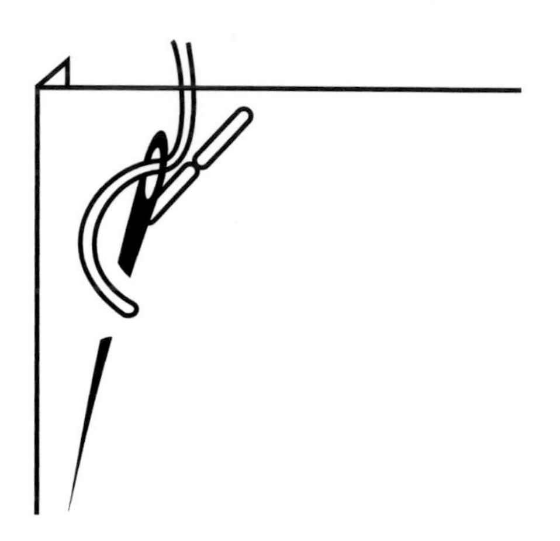

THE CROSS STITCH

This is a great stitch for adding color and texture to embroidery projects. Form X shapes on one side of the fabric by starting with a series of parallel angled stitches; then turn the fabric around and stitch back the other way to finish the Xs.

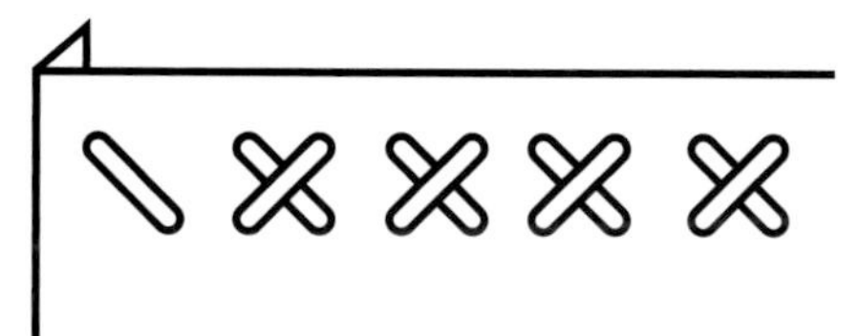

Sewing Circuits: Tools and Materials

Figure 1.4 shows the tools and materials that are more specific to sewing circuits. You probably won't find most of them in craft stores, so you'll have to order them online. You can find a list of recommended suppliers at the website for this book: www.techdiy.org.

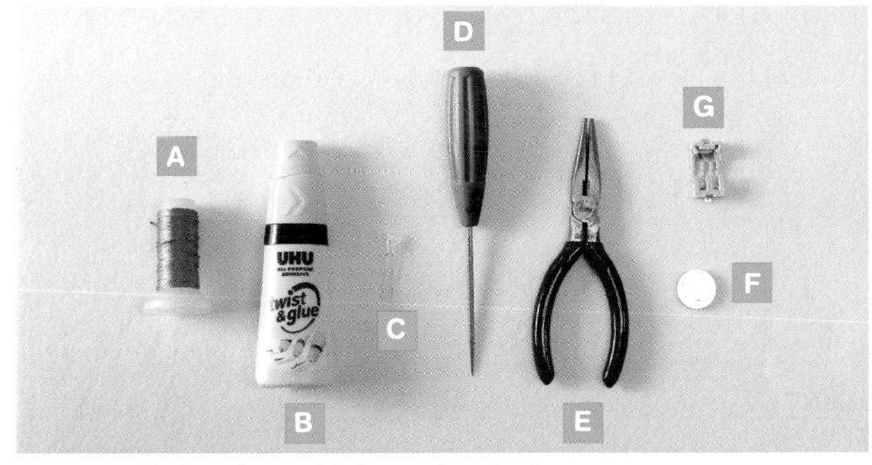

FIGURE 1.4: Tools and materials for sewing circuits

A **Conductive thread**—Ordinarily electronic circuits are made with wires and metal. However, the projects in this book use conductive thread. You can purchase conductive thread from online stores like adafruit.com or sparkfun.com. Some craft stores carry metallic thread; please notes that this thread is *not* conductive and won't work for sewing circuits.

B **Glue**—You can use glue to attach battery holders or other objects onto felt sheets. Glue is also good for keeping conductive thread knots from coming undone. We recommend a fast-drying fabric glue.

C **LED**—LEDs, or light-emitting diodes, come in several colors and sizes. For the projects in this book, we use LEDs that are 5 mm (millimeters) in diameter. You can choose whatever colors you like.

D **Awl**—We use a sewing awl to poke holes in felt sheets so that we can insert LED legs or other electronic components. Any pointy, sharp tool will work, but be careful when poking holes in fabric so you don't accidentally poke yourself!

E **Needle-nose pliers**—Pliers make it easier to connect conductive threads to LEDs or other electronic components. We will use the pliers to wrap the legs of the LEDs into coils. Needle-nose, or long-nose, pliers are best.

F **Battery**—The projects in this book use small, three-volt (3V) coin cell batteries. This type of battery is commonly called a CR2032. These are easy to find at most stores that carry batteries.

G **Battery holder**—The battery holders for these batteries are a bit harder to source. For the projects in this book, you want to find a battery holder that is easy to sew with. We recommend this one: https://www.adafruit.com/products/653. These holders have small holes you can sew through. You'll have to use a #5 sewing needle or smaller to get through them.

How to Use This Book

All of the chapters in this book are split into three sections: Learn, Make, and Explore.

STEP 1: LEARN

You'll start each project by learning about how the specific circuit focused on in that chapter works. Each chapter introduces a new electronic component or concept. Try to understand how electricity is going to flow through your circuit and the attached electronic components before you start sewing. As you continue to work through these projects and progress through the book, you will learn about some more advanced concepts of electricity and how to work with many different electronic components.

STEP 2: MAKE

Next you will start by creating your own ideas on how to use the circuits in each chapter. Don't try to copy all the projects in this book exactly. We recommend that you change the ideas a bit to make them more fun and relevant to your interests. As you come up with new and increasingly complicated ideas, don't try anything that seems too difficult before you are ready. Attempt to find a balance between pushing the boundaries of your skills and continuing to build on the knowledge that you learn from the projects in this book.

Always start designing your projects and circuits using paper and pencil and check carefully for any possible problems before you begin sewing. Sewing circuits can be difficult. It's a lot easier to fix problems if you catch them before you sew them down.

Make your projects using the guides in each chapter. Follow the steps in the project guides for sewing your circuits. Be especially careful that you don't accidentally tangle, cross, or cut your conductive thread between components.

STEP 3: EXPLORE

The final section of each chapter presents some more project ideas that use the concepts you learned in that chapter.

After you finish each project, it is a great idea to reflect and write a little bit about your experience. What did you learn by making this project? What would you do differently if you made another version? If you take a photograph and post it onto a website with your reflections, for instance, other people might learn from your project! This is a wonderful opportunity to start and share a portfolio of your own craft and technology projects.

We recommend doing the projects in this book in order, starting with the simple circuit in the Happy House Project in the following chapter and culminating in the final project in Chapter 10, which involves computer programming. Each project is meant to introduce a new concept or electronic component, and each chapter builds on the skills you learned in previous chapters. Although the subtitle of the book is *Easy Electronics Projects for Parents and Kids*, some of the later activities, especially the projects in Chapters 7, 8, and 9, are fairly challenging. But as long as you work your way through the first few projects, you will learn the skills you need to accomplish those more complex projects.

Have fun exploring the wonderful world of electricity and creating your own projects that light up, move around, and make sounds!

Chapter

2

My Happy House

MAKE A CIRCUIT

In this introductory project, you learn about electricity, circuits, and some basic electronic components. You will start your first project by drawing a simple picture on paper. Then you will embroider that drawing onto a sheet of felt. Finally, you'll connect an LED (light-emitting diode) to a battery with conductive thread to create a simple light-up circuit on your embroidered picture.

In this section, you will learn about some common electronic components and make your first circuit. But first you need to learn a little bit about what electricity is and how it flows through a circuit.

How Does Electricity Flow?

When you flip the switch to turn on the lights in your home, you close a circuit between a power source and your lights that allows electricity to flow. Usually circuits are made from copper or other metals because these materials are good at conducting electricity. When you connect a circuit to a power source such as a battery, electrons leave the negative side of the battery and push free electrons in the copper toward the positive side of the battery, like balls being pushed through a tube. Each individual electron actually moves really slowly—about 1 meter per second. But because the electrons are all pushing each other through the circuit, the lights turn on right away.

Light-Emitting Diodes (LEDs)

An LED is a small, energy-efficient light bulb that has many applications. You'll find LEDs in everything from toys to real spaceships.

LEDs have two wire "legs." The longer leg connects to the positive side of a battery whereas the shorter leg connects to the negative side. You will curl the legs of your LEDs to make it easier to sew with them, so it will be difficult to see the differences in length when you are working. Here are two other ways to determine which leg is positive and which is negative.

+ −

Examine the bottom of the LED where the wires come out. One side of the plastic is flat. This is the negative side.

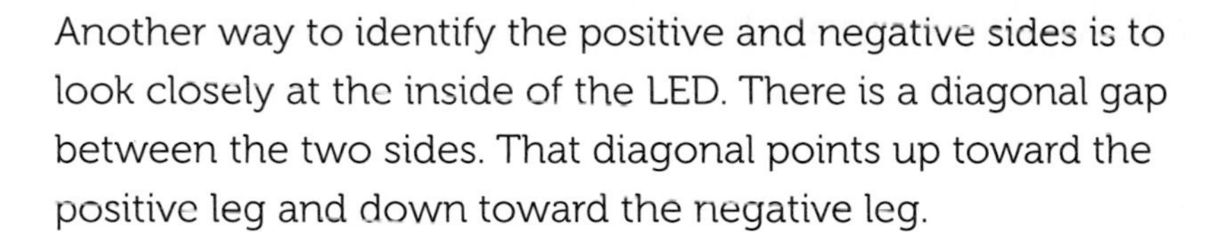

Another way to identify the positive and negative sides is to look closely at the inside of the LED. There is a diagonal gap between the two sides. That diagonal points up toward the positive leg and down toward the negative leg.

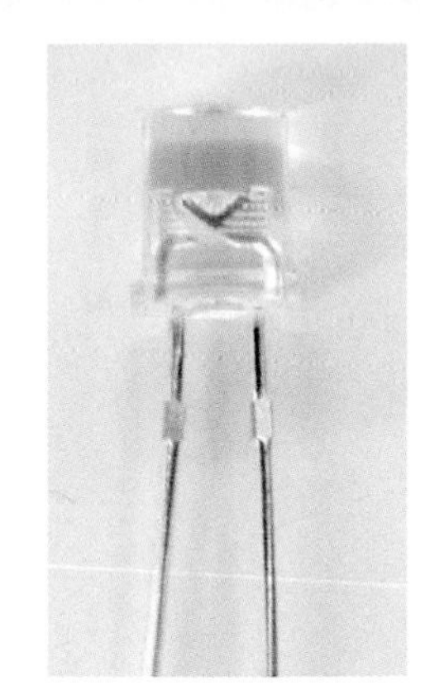

Battery

In order to turn on an LED, you need at least 3 volts of electricity. Since LEDs are energy efficient, they can last a long time—more than 10 years! The projects in this book use a 3V (three-volt) lithium cell battery called a CR2032 coin cell battery. These batteries have a + symbol on the positive side. The other side is the negative side. They are fairly common and easy to find in stores that sell batteries.

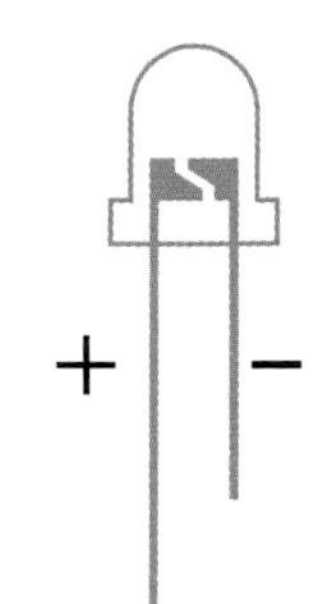

Battery Holders

In order to easily connect and remove batteries from your circuits, you need to use a battery holder that is simple to sew. I recommend this battery holder because it has holes that you can pass a needle through. You can find such a battery holder at Sparkfun (www.sparkfun.com) or Adafruit (www.adafruit.com).

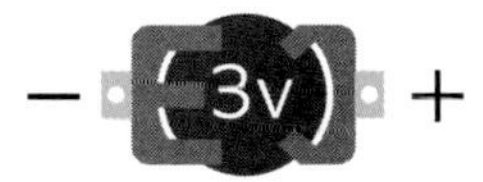

Let There Be Light!

To test your battery and LED, put the battery between the two legs of the LED. Make sure the long leg touches the positive side of the battery and the short leg is connected to the negative side.

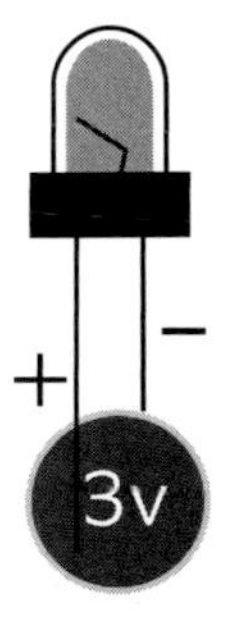

CAUTION: Some LEDs can be pretty bright, especially if you look at them from the top down. Be careful not to stare at the LED too closely. It can damage your eyesight!

Symbols of Electric Circuit Diagrams

Take a look at Figure 2.1. Electronic circuit diagrams use these symbols for electronic parts. If you can remember these symbols, you'll be able to both read and draw diagrams like the one in Figure 2.2.

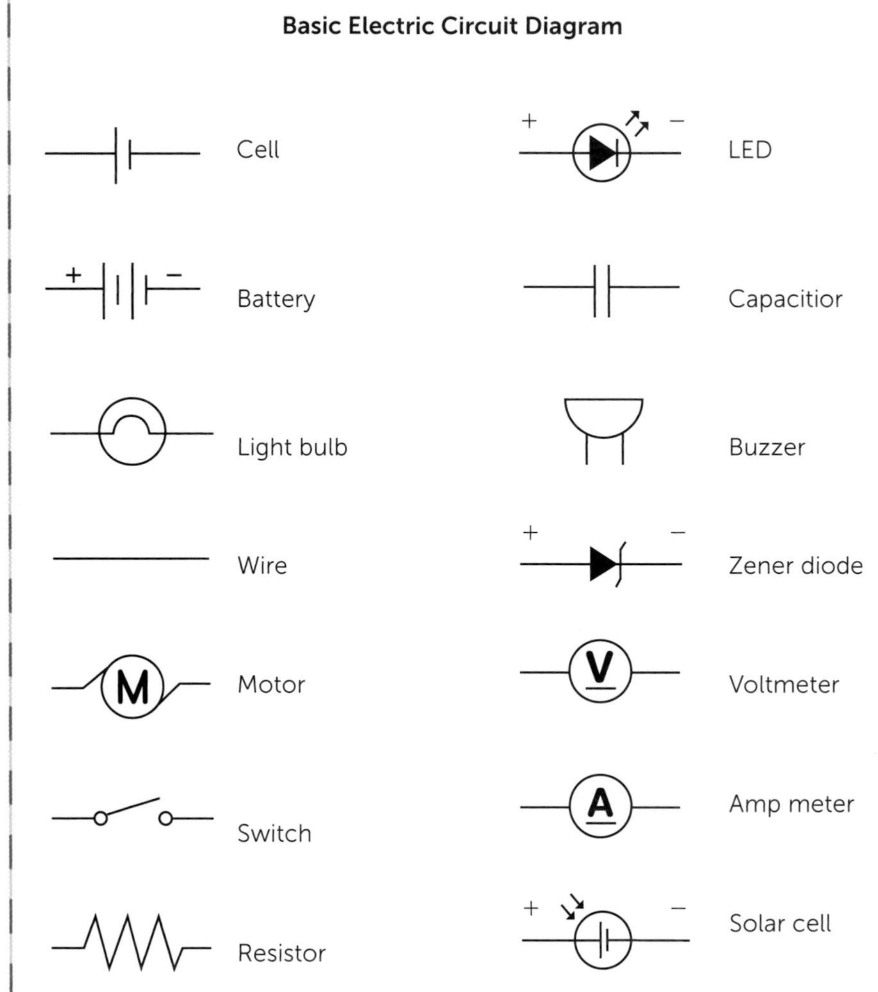

FIGURE 2.1: Symbols to use for electric circuit diagrams

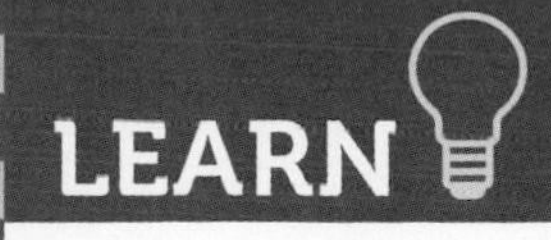

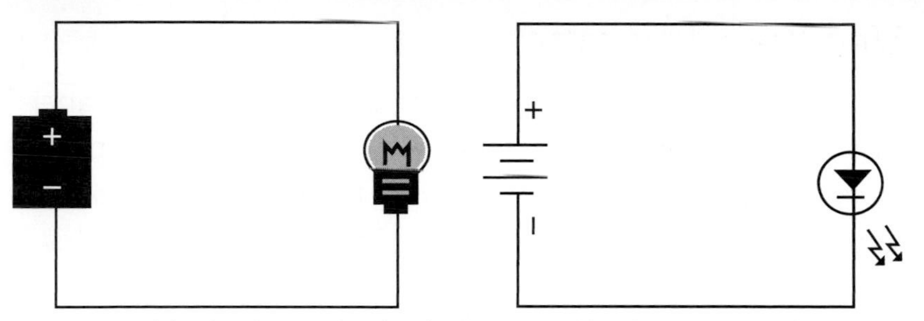

FIGURE 2.2: A basic electronic circuit diagram with a battery and a light

Use Your Imagination

Start by drawing a sketch on paper. Choose a piece of paper that is the same size as your felt or fabric sheet. Make sure to leave enough space in your drawing for your LED and battery holder. Try to come up with a sketch that cleverly uses the circuit and needlework together as in Figure 2.3.

FIGURE 2.3: Sketch your design on paper.

Plan It Out

Now that you've come up with an idea of what you want your project to look like, draw in your circuit with the LED and battery holder and connect it to the rest of your drawing with dashed lines (see Figure 2.4). Make sure to label the positive and negative side of your LED and battery holder with the appropriate symbols. Remember, the long leg of the LED needs to be connected to the positive side of the battery and the short leg must connect to the negative side.

CAUTION: These two dashed lines should never cross or touch each other. This would cause a short circuit, cause the battery to drain quickly, and make it so your LED would not light up.

FIGURE 2.4: Drawing with LED and battery holder included

Preparation

Now that you have an idea and a sketch for your project, you can start to sew the image and circuit onto fabric. Before you begin, make sure that you have all the supplies that you need. When you are sewing, take your time and make sure that each stitch is neat and complete before you begin the next stitch. It is okay if you make a mistake or get tangled. Just take out that thread and start the section over again. You will get better over time! Feel free to refer to the "Basic Needlework: Knots and Stitches" section in Chapter 1 for tips and help.

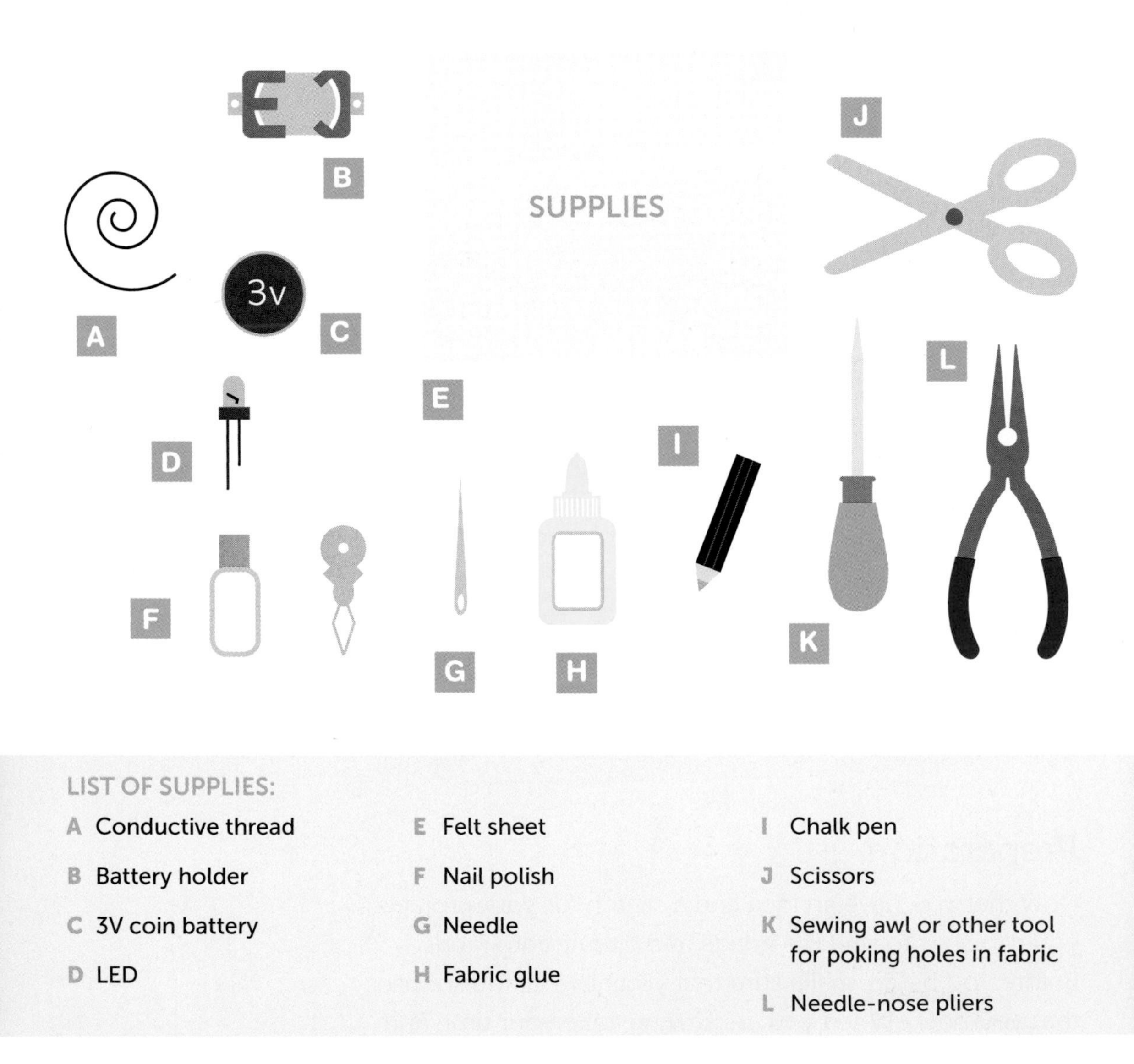

LIST OF SUPPLIES:

- A Conductive thread
- B Battery holder
- C 3V coin battery
- D LED
- E Felt sheet
- F Nail polish
- G Needle
- H Fabric glue
- I Chalk pen
- J Scissors
- K Sewing awl or other tool for poking holes in fabric
- L Needle-nose pliers

Make It

Follow these steps to make your first soft circuit project:

1. Based on your paper sketch, use a chalk pen to copy your drawing onto a sheet of felt or fabric.

2. Use colorful embroidery thread to make running stitches, back stitches, and cross stitches to sew the drawing onto your fabric. If sewing and embroidery are new to you, it might be easier to use an embroidery hoop to hold the felt or fabric while you are working.

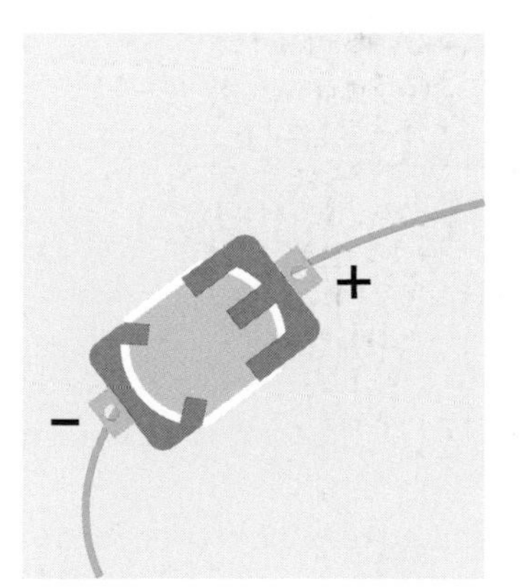

3. Check the location of the battery holder's positive and negative sides. Make sure to place it in the correct orientation and then paste it to the fabric using fabric glue. Let the glue dry. Gluing the battery holder to the fabric is optional, but doing so will make it easier for you to sew your circuit.

4. Poke holes for the two legs of the LED using an awl. Note which hole is for the positive leg and which hole is for the negative leg.

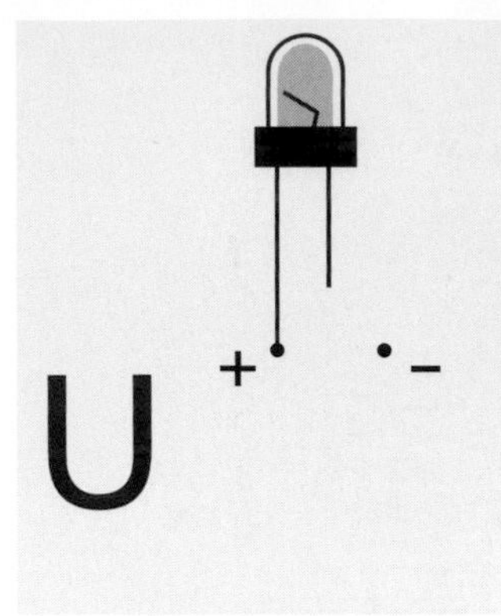

5. Check the length of the LED legs. Make sure to put the long leg in the hole that is pointed toward the positive side of the battery holder; put the shorter leg in the hole that points toward the negative side. Using pliers, curl the legs of the LED on the underside of the fabric into a spiral to which you can attach the conductive thread.

6. Using conductive thread, sew the LED's positive leg to the positive side of the battery holder. Do the same with the negative leg and the negative side of the battery holder.

7. Insert your battery into the holder. Make sure to place it correctly—the positive side of the battery should face up. If you've succeeded, your LED should light up!

8. Conductive thread tends to fray and can become unknotted where you connect it to the LED and battery holder. To prevent this from happening, you can use a small dab of fabric glue or clear nail polish on those four spots.

9. Now that you're finished, you can place your soft circuit in a frame so you can hang it up. Just take out the battery when you want to turn it off.

In Chapter 3, you will learn how to add a switch so that you can turn your circuits on and off.

My daughter inspired this project. During a visit to a children's museum in New York City, she made this painting, called "The place where I live," using watercolors.

Here are some other ideas for practice sewing and basic electronic circuits using cute pictures that include LED lights!

Hello Snail

As we explained earlier, when you are adding circuits to your sewing projects, you cannot allow the positive and negative sides of the conductive thread to cross. This may seem like a limitation; however, sometimes constraints become sources for ideas. This is how I had the idea for the snail circuit in Figure 2.5. Doesn't the battery holder work well as part of the snail's house?

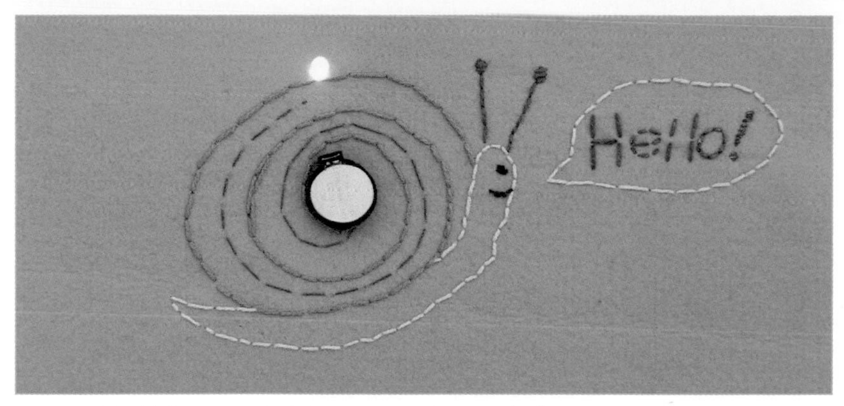

FIGURE 2.5: Using constraints as part of your design

Other Basic Electronic Circuit Projects

In addition to creating your own circuit projects, you can add circuits to existing needlework. For instance, take a look at this embroidered tablecloth with LEDs sewn into the center of a flower.

A Headband with Ornaments and Glowing Lights

You might want to try making a headband like this one my daughter and I made; you can add many interesting ornaments, like pins and buttons. If you add LEDs, like we did, they just seem to blend into the other ornaments! I found that this headband's twinkling lights made it easy to spot my daughter on the playground in the evening!

What's Next?

Great job on sewing your first circuit! To turn the circuit off, you have remove the battery. Wouldn't it be nice if you could turn it on and off with a switch? In the next chapter you are going to learn about what types of material conduct electricity and what types do not. You will also learn how you can use this information to control the flow of electricity in a circuit using a homemade switch.

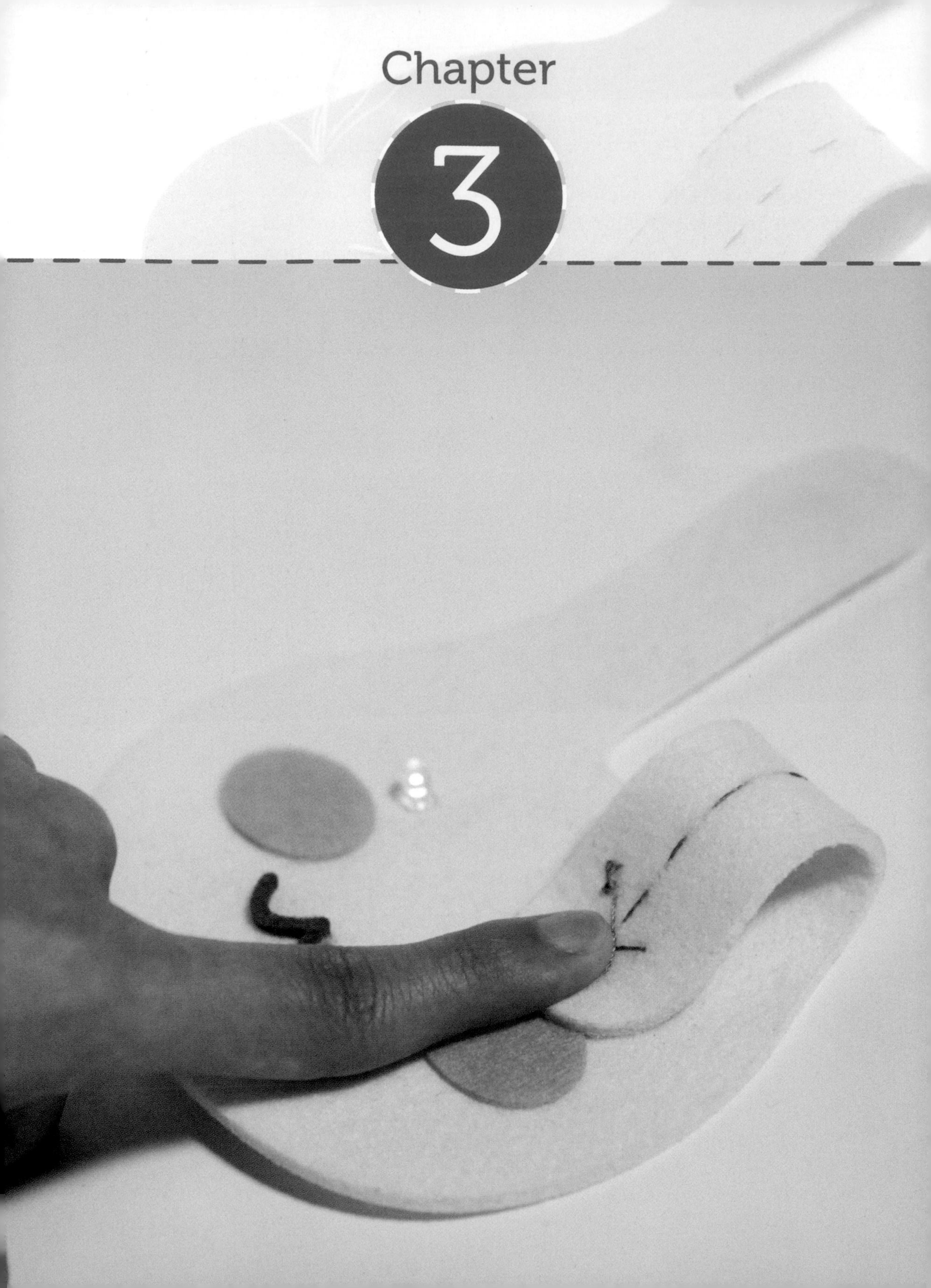

Chapter

3

A Winking Rabbit Doll

CREATE A SWITCH

In the last project, you made a simple circuit that you sewed onto fabric, but the only way to turn off your circuit was to remove the battery. Wouldn't it be better if you had an easy way to turn your project on or off? In this chapter, you are going to learn how to add on/off switches to your circuits. You will learn a bit about different types of switches and buttons. You will also learn why electricity flows so well through some materials but not through others. Figure 3.1 shows the sample project you will be following in this chapter.

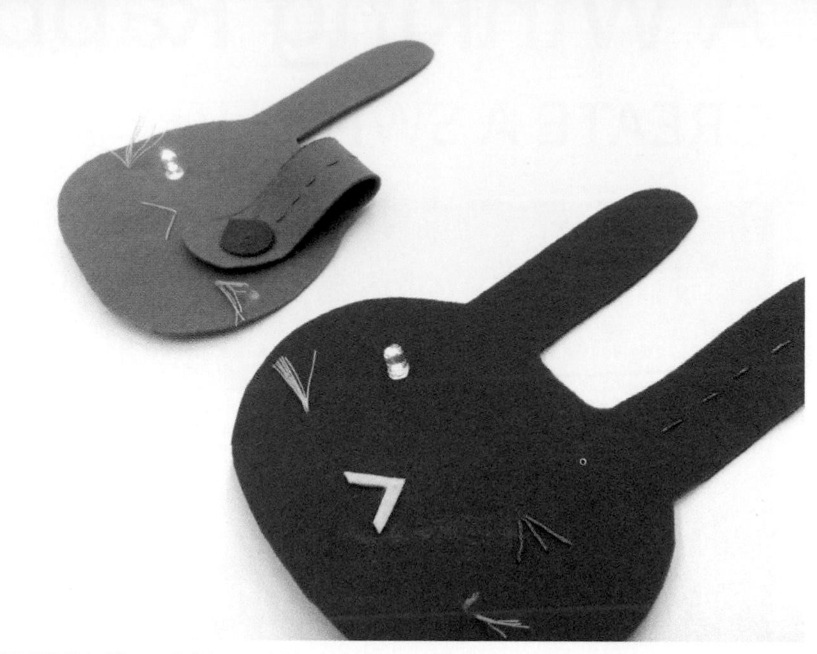

FIGURE 3.1: The winking rabbit. When you fold the right ear down to touch its eye, the LED lights up.

As long as a circuit has a closed path through which electricity can flow from the negative side of a battery to the positive side, the current flows without stopping until the battery runs out of charge. But the moment that the path is broken, cut off, or opened, the electricity stops flowing. This is what happens when you turn off the lights in your home. When you flip the switch to the off position, a gap, or break, forms in the circuit, preventing the electricity from flowing. A *switch* is simply a tool that allows us to open and close a circuit. Figures 3.2 and 3.3 show two common types of switch and their electronic symbols.

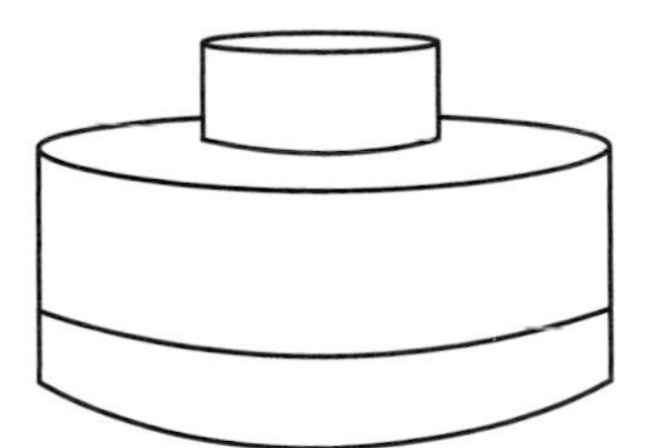

FIGURE 3.2: A button

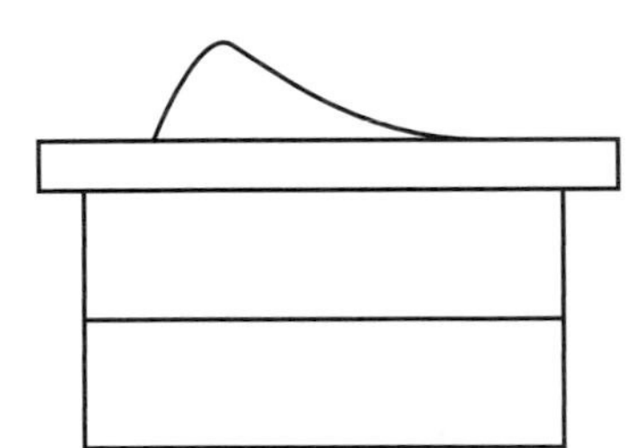

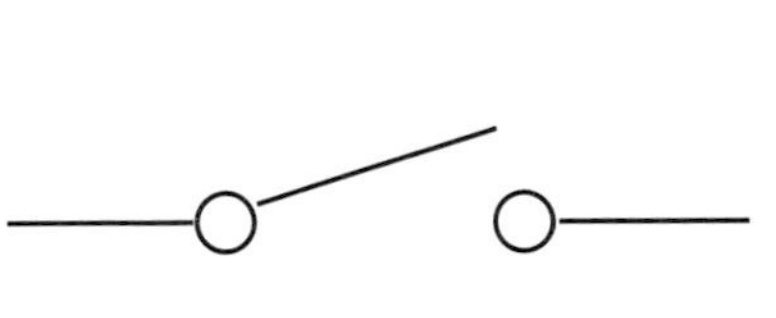

FIGURE 3.3: A toggle switch

Conductors and Insulators

All material is made up of basic building blocks called atoms. Every atom of material has at least one electron moving around it. Some materials, such as metals, have free electrons that can move from atom to atom. This movement of electrons from atom to atom is what carries electricity through a circuit. Materials with free electrons allow electric currents to flow through them. These materials are called *conductors*. Gold, silver, copper, aluminum, and iron are examples of good conductors. Materials that do not have free electrons do not allow electricity to flow through them. These materials are called *insulators*. Glass, rubber, wood, paper, and plastic are insulators. By using a combination of conducting and insulating materials, we can make our own switches.

The wires that you see in electronic devices are made up of a conductive material, usually copper, that is covered with an insulating material like plastic or rubber (see Figure 3.4). Electricity can easily flow through a wire, but if two wires cross, their insulation prevents short circuits. Conductive thread does not have insulation, so we have to be extra careful not to let two threads touch or cross each other in our circuits unless they are supposed to cross.

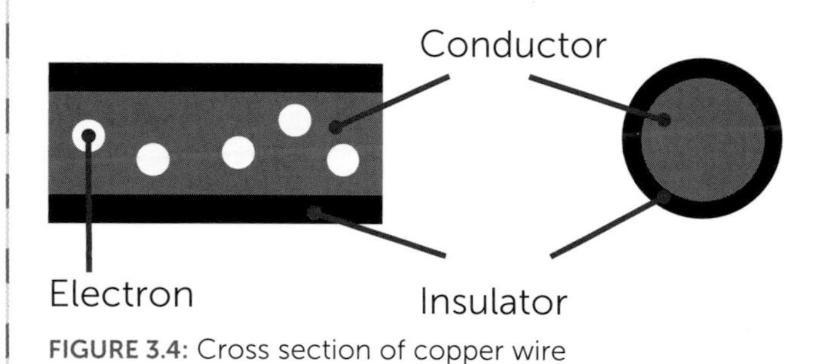

FIGURE 3.4: Cross section of copper wire

In the Winking Rabbit project, you will make a fabric switch out of conductive thread and felt. However, you can also make switches with snaps, pins, and conductive fabric.

The circuit diagram in Figure 3.5 is the same as the one from Chapter 2, except this one includes a switch. In this diagram, the switch is open, so electricity will not flow through the circuit, but as soon as you close the switch, the light emitting diode (LED) will turn on.

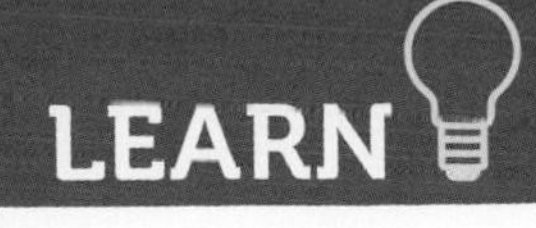

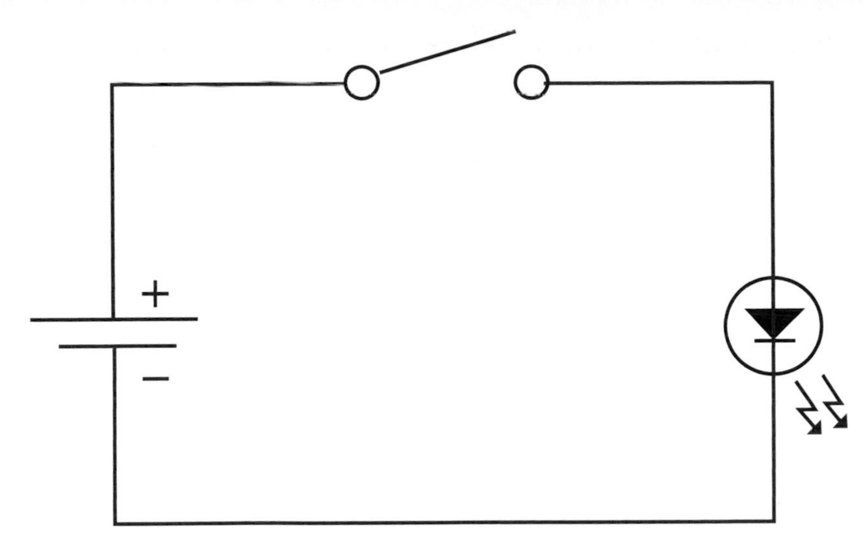

FIGURE 3.5: Basic circuit diagram including a battery, an LED, and an open switch

Use Your Imagination

Now that you've gotten used to the ideas of circuits and switches and what materials you can use to make them, take a minute to imagine a picture or character that includes a circuit with an LED and a battery. This time, when you're thinking about how to design your circuit, think of a place where you can leave the circuit open, but also where you can include a switch that allows you to close the circuit and turn an LED on and off, as I have done in Figure 3.6. If you are looking for more ideas, check the "Explore" section at the end of this chapter.

Plan It Out

Now that you've sketched your project, it's time to draw your circuit, including the LED and battery. Make sure you mark the positive and negative sides of your LED and battery

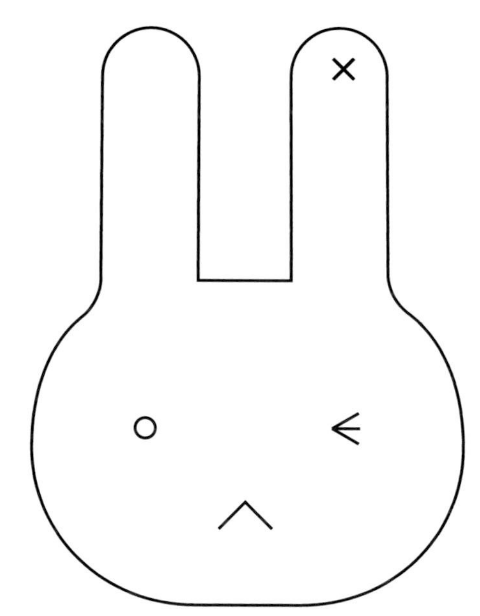

FIGURE 3.6: The front of the winking rabbit. The LED is in the left eye. When you bend the right ear down to touch the right eye, a switch closes and the LED lights up.

so that you don't get confused. *Remember that the positive and negative sides of your circuit can never cross.* Also, make sure to include a spot for your switch, which will open and close your circuit. You can include the switch on either the positive or negative side of the circuit. To see an example, take a look at my drawing of the winking rabbit circuit (see Figure 3.7). When you fold the right ear down (which is actually on the left side in Figure 3.7 because you're looking at the back side of the rabbit) and touch it to her right eye, the circuit completes and the LED in the left eye lights up!

Preparation

After you are done sketching your project and adding a plan for your circuit with a switch, you can start to cut out your character and sew the circuit onto the fabric. Take your time and try to keep your stitches as neat as possible. Make

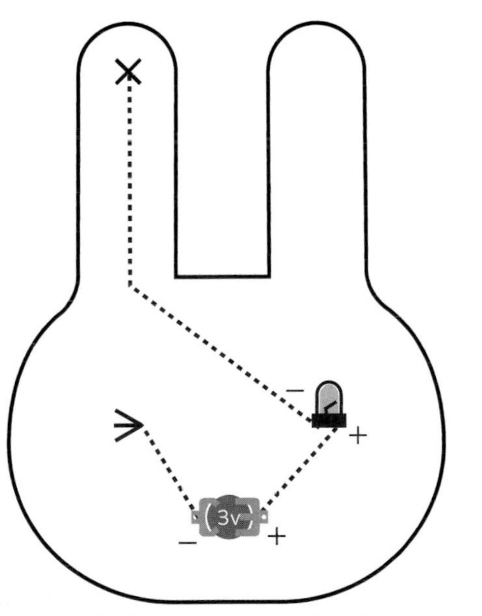

FIGURE 3.7: This is the back of the winking rabbit. It reveals the circuit with the open switch.

sure each stitch is complete before you start the next stitch. You can always refer to the "Basic Needlework: Knots and Stitches" section in Chapter 1 for tips and help.

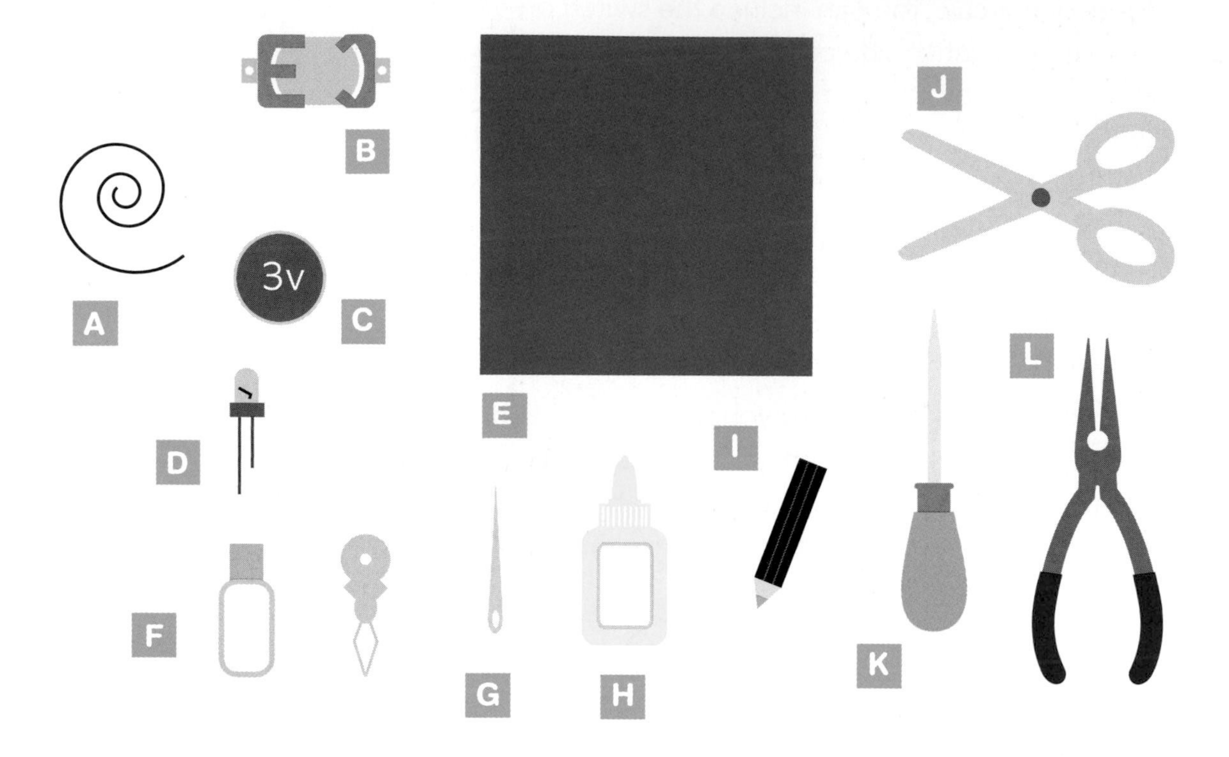

LIST OF SUPPLIES:

- A Conductive thread
- B Battery holder
- C 3V coin cell battery
- D LED
- E Felt sheet
- F Nail polish
- G Needle
- H Fabric glue
- I Chalk pen
- J Scissors
- K Sewing awl or other tool to poke holes in fabric
- L Needle-nose pliers

Make It

Follow these steps to make your second soft circuit project:

1. Based on your paper sketch, use a chalk pen to draw the outline of your project onto fabric and then cut it out with scissors.

2. On the front of your project, mark the spot for the LED (the rabbit's left eye) and the two connection points for your switch (the rabbit's right eye and the top of the right ear).

3. On the back of your project, use a chalk pen to sketch out your circuit. Make sure to include places for your LED, battery holder, and the two connection points for your switch. Also mark the positions of the positive (long) leg of your LED and the positive side of your battery holder by making plus signs (+) on the fabric.

4. Turn the project back to the front side. Using an awl, in the spot where you are going to put your LED, poke two holes right next to each other in the felt.

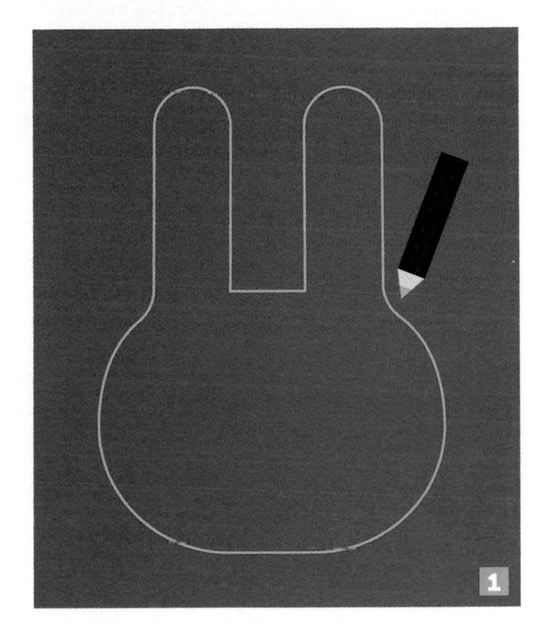

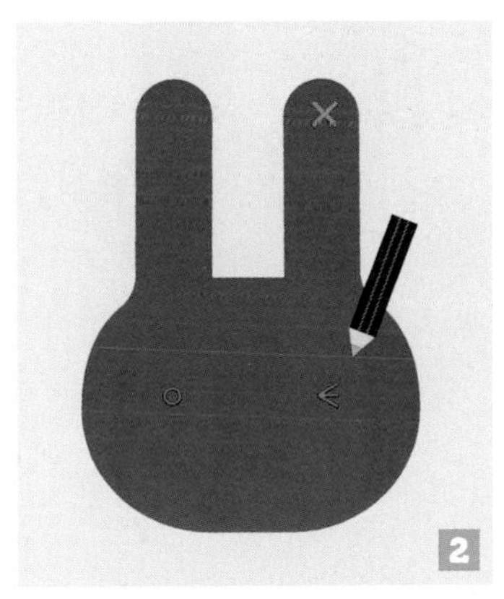

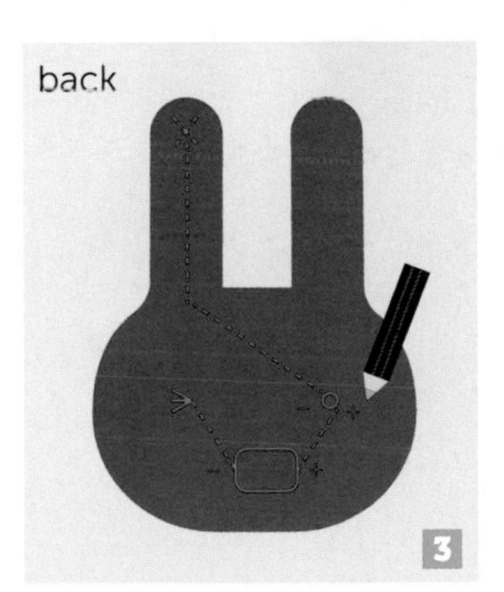

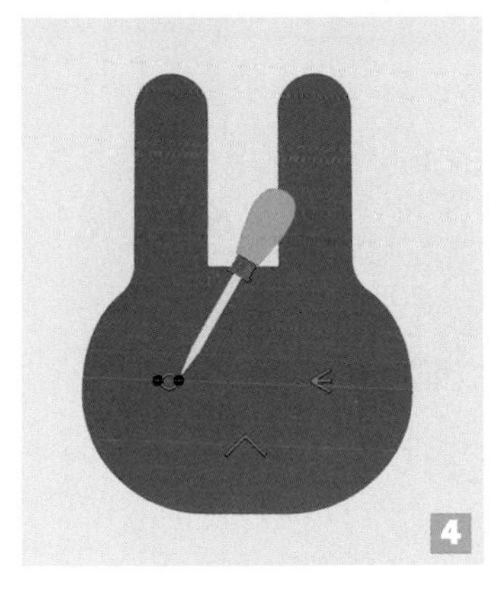

5. From the front of the design, push each leg of the LED through the two holes in the felt. Make sure to put the positive (longer) leg of the LED into the hole that you marked positive (+) in step 3.

6. Flip the project over to the back. Using needle-nose pliers, curl up the legs of the LEDs so that it is easier to sew to them. Remember which leg is the positive leg and which is the negative leg! *Note that in this illustration, the LED is shown on the back for clarity.*

7. Using fabric glue, glue down the battery holder to the back of your fabric. *Note that in this illustration, the LED is shown on the back for clarity.*

8. Using conductive thread, sew the positive leg of the LED to the positive side of the battery holder. *Note that in this illustration, the LED is shown on the back for clarity.*

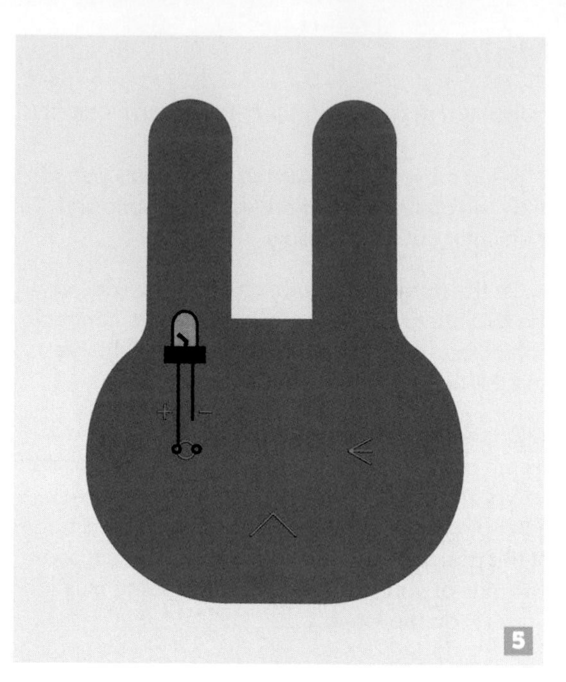

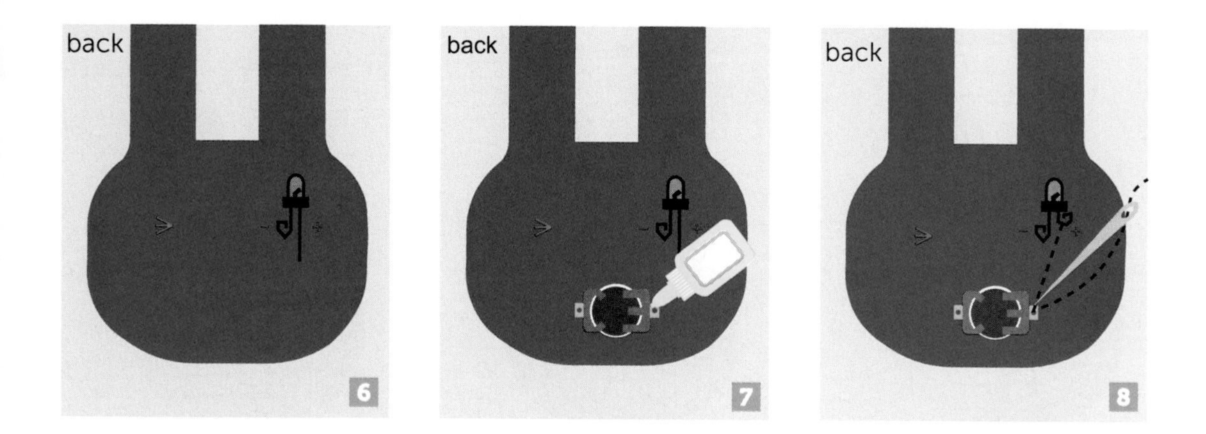

9. Using conductive thread, sew the negative side of your LED to the farther of the two connection points on your switch—the rabbit's ear). *Note that in this illustration, the LED is shown on the back for clarity.*

10. Using conductive thread, sew the negative side of your battery holder to the other connection point on your switch (the right eye). *Note that in this illustration the LED is shown on the back for clarity.*

11. Add a battery to your battery holder.

12. Flip your project back over and test your circuit by folding down the right ear to connect the two ends of your switch and close the circuit. If the light turns on, your project is done!

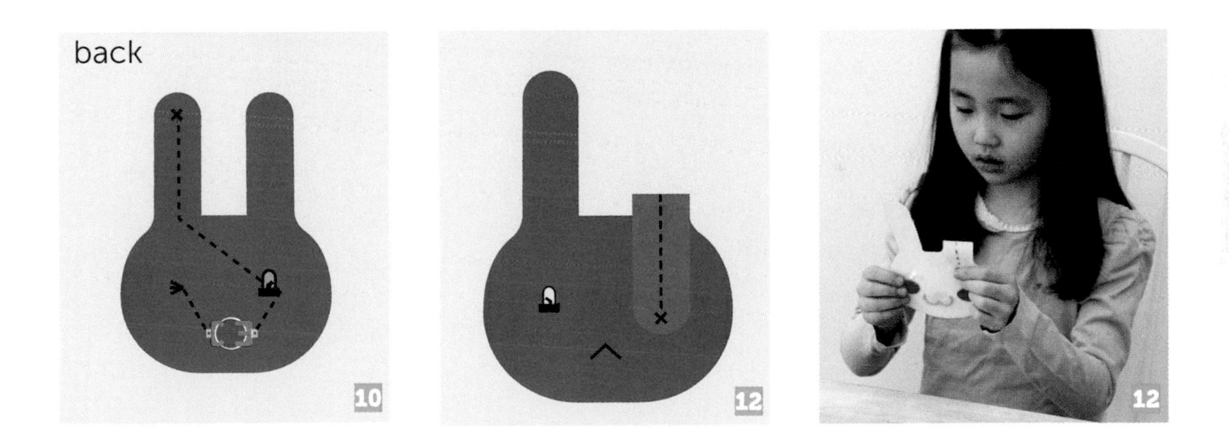

When you are finished, the back of your circuit will look something like Figure 3.8.

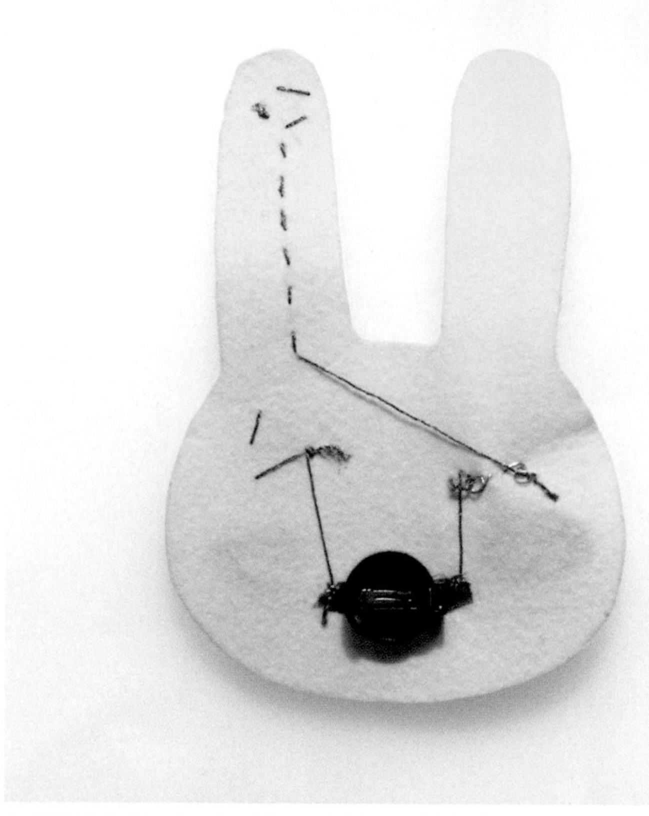

FIGURE 3.8: The back of the winking rabbit circuit

Figures 3.9 through 3.11 show a few other fabric projects that you can make with switches. Try some out!

You can make your own switches out of many different materials that conduct electricity, including tinfoil, safety pins, and conductive fabric. I'll now describe a project that uses metal snap fasteners to turn on and off.

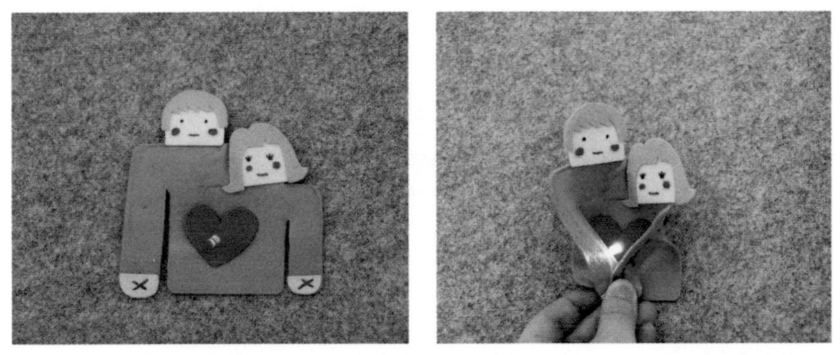

FIGURE 3.9: These two turn on their love and then become one by holding hands.

FIGURE 3.10: Rudolph's nose gets red when his bell is pressed.

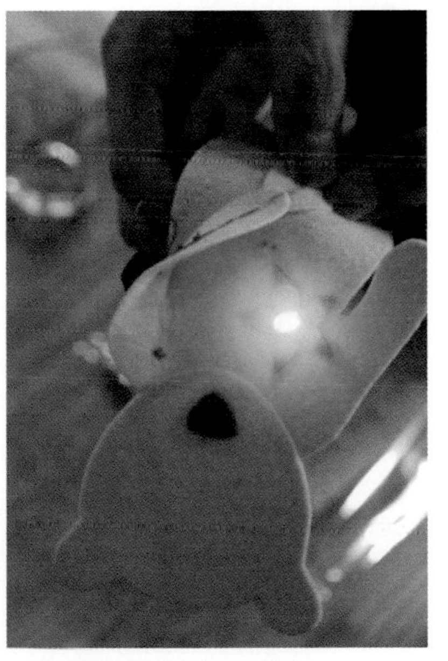

FIGURE 3.11: This bear lights up when he touches his belly.

Using Snaps to Make a Wearable Switch

Before I started using felt in my projects, I experimented with different materials. One project, called the "Secret Tree," used LEDs and parchment paper to make light-up flowers (see Figure 3.12).

I loved the way that the combination of the material, form, and LED made it look like a flower in the moonlight. This project led me to start experimenting with fabrics and felt.

I then made several versions of the moonlit flower in white felt. I used needlework to crease the felt and make the flowers pop in three dimensions.

Eventually I figured out how to make a wearable ring with the moonlit flower that you turn on and off using a metal snap button as a switch. You can see examples of these projects in Figures 3.13–3.16. If you'd like to make these flowers yourself, see the Appendix for instructions.

FIGURE 3.12: Parchment flower with LED

FIGURE 3.13: A cluster of fabric moonlit flowers

FIGURE 3.14: A moonlit flower in felt

FIGURE 3.15: The moonlit flower uses metal snaps as the switch.

What's Next?

Now that you know why some materials conduct electricity and some don't and how to use conductors and insulators to make a switch, the next chapter will build on this circuit to add a second LED.

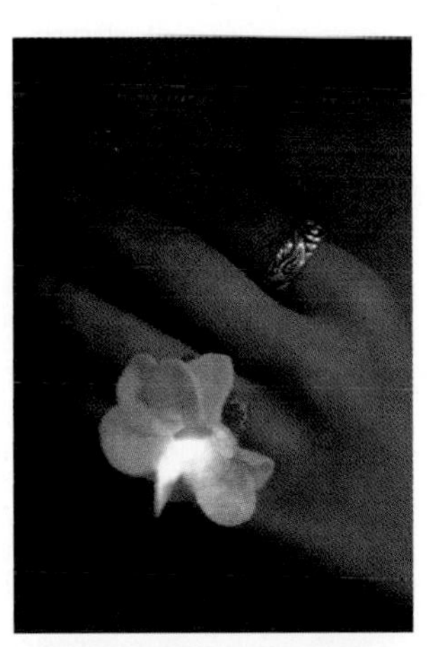

FIGURE 3.16: Wearing the moonlit flower as a ring

Chapter
4

A Silly Ghost

WORKING WITH PARALLEL CIRCUITS

In this chapter, you are going to learn about two different kinds of circuits: series circuits and parallel circuits. Then you will make a project that uses both a simple circuit and a switch; you will also add a second LED to your circuit.

When you press down the ghost character's tongue, both his eyes light up! Do you think you know how it works?

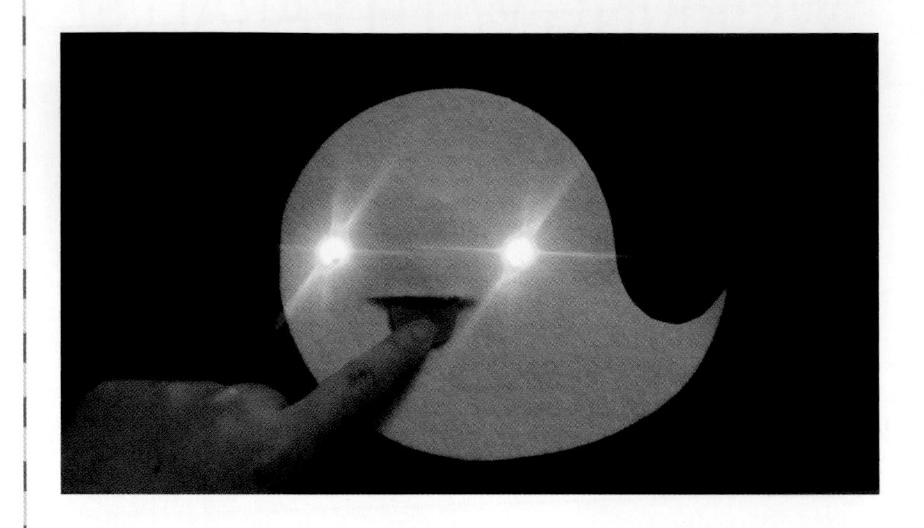

This is very similar to our last project; the only real difference is that we added a second LED. You can arrange this type of circuit in two ways: as a series circuit or as a parallel circuit.

Series Circuit

In a series circuit, the components (LEDs, switches, etc.) are arranged so that the electric current passes through them in a line, or a series (see Figure 4.1). The electricity can follow only one path in a series circuit. Current passes from the power source, through each component, and back to the power source. If one of your components gets damaged, the electricity won't flow through the circuit anymore, and it stops working.

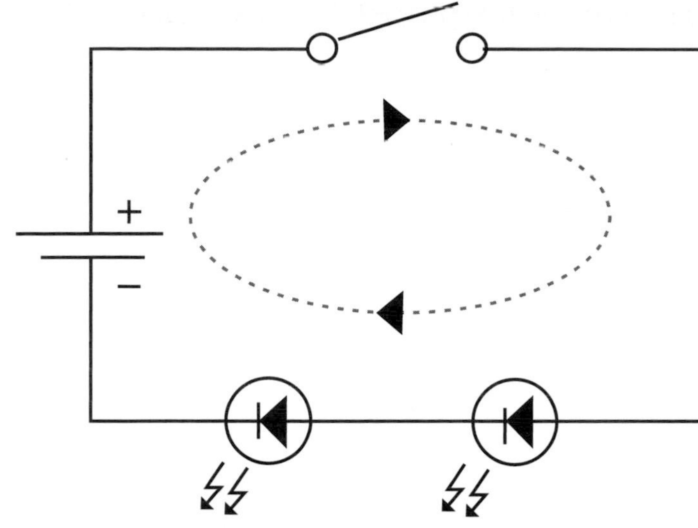

FIGURE 4.1: A series circuit with two LEDs and an open switch

Parallel Circuit

In a parallel circuit (see Figure 4.2), the components are arranged so that at least two paths can carry the electric current. If both paths have the same components, then the electrical current is divided evenly between the two paths. If one of the components gets damaged, the electricity still has another path the current can flow through.

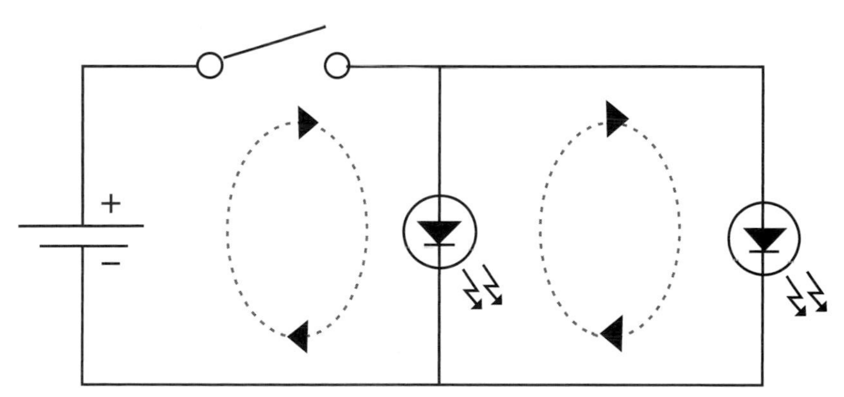

FIGURE 4.2: A parallel circuit with two LEDs and an open switch

Real-World Examples of Series and Parallel Circuits

If you've ever had a string of Christmas lights that goes dark when one of the light bulbs is removed or burns out, then you've used a series circuit. But if all the appliances in your home were connected in a series circuit, then you'd have to turn on your toaster in order to run the microwave. That is why the electrical circuits in our homes are mostly set up in parallel.

The ghost circuit you build in this chapter uses a parallel circuit and a switch, as shown in Figure 4.3. When you press the ghost's tongue, it will act as a switch that turns on the two LEDs in its eyes.

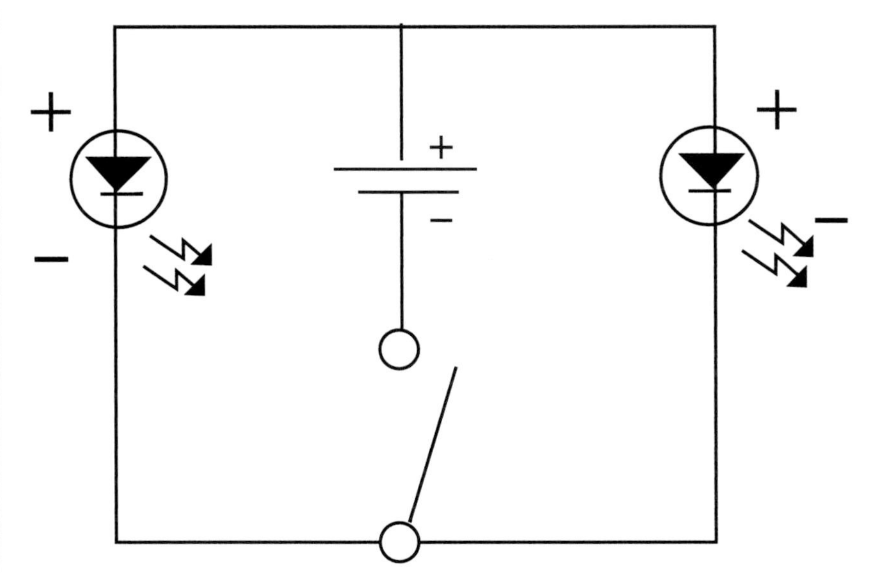

FIGURE 4.3: A parallel circuit with two LEDs and a switch. This arrangement of the parallel circuit is better than the one in Figure 4.2 for the Silly Ghost project.

Use Your Imagination

Now let's get to work. Think of a fun idea that uses a switch to turn on two lights. Maybe it's a doll that looks like you, or a cartoon character that you like. Make a sketch and think about where you can put the battery, the lights, and the switch. Here is a sketch of the ghost example in this chapter. The LEDs go in the eyes and the switch is under the tongue.

Plan It Out

Now add a drawing of your circuit to the back of your sketch like the one in Figure 4.4. Make sure to include two LEDs in parallel. Both of the positive legs of the LEDs should be connected to the positive side of the battery. Both of the negative legs of the LEDs should be connected to the negative side of the battery when you close your switch. Be sure to make all your positive and negative connections in your circuit. Figure out where you want to put your switch. In the Silly Ghost, the switch is located under the ghost's tongue and mouth.

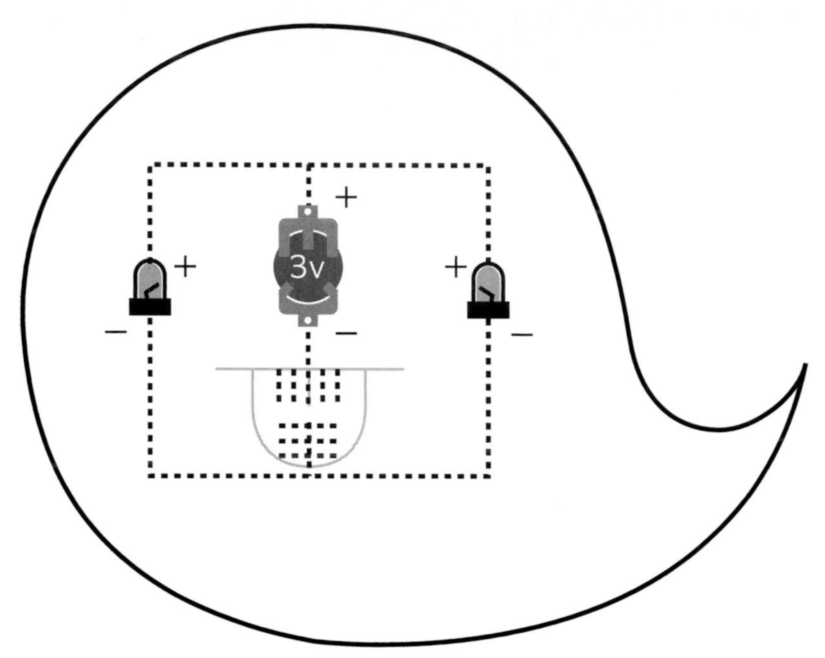

FIGURE 4.4: Sketch of a parallel circuit on the back of the Silly Ghost. Includes two LEDs, a battery holder and a switch under the tongue.

Preparation

After you are done sketching your project and a plan for your parallel circuit with a switch, you will cut out your character and sew the circuit onto fabric. This circuit is a bit more complicated than the last circuit. Remember to take your time and try to keep your stitches as neat as possible. You can always refer to the "Basic Needlework: Knots and Stitches" section in Chapter 1 for tips and help. It is possible to hide your conductive thread stitches so that you can't

see the circuit from the front, like I did in the ghost project. When you poke the needle into the back of the project, do it at a shallow angle and try not to poke all the way through the felt before you poke the needle out of the back of the felt again. This technique is easier with thicker felts and after a bit of practice!

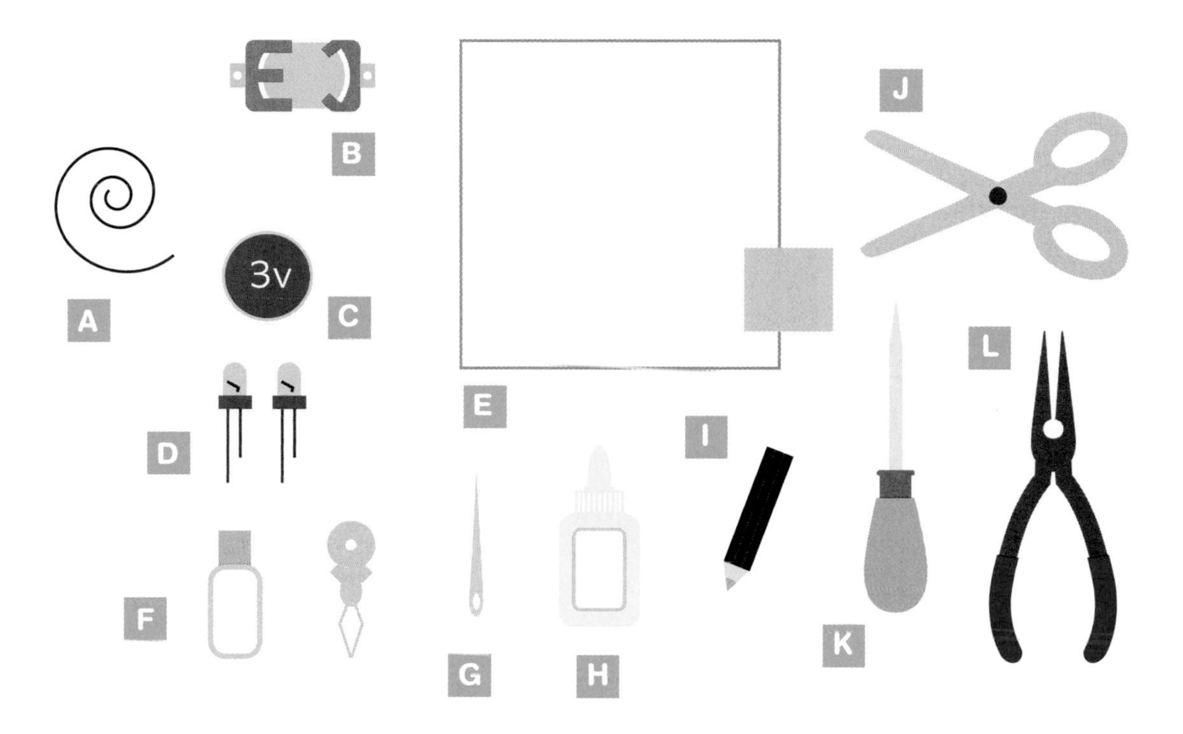

LIST OF SUPPLIES:

A Conductive thread

B Battery holder

C 3V coin battery

D 2 LEDs

E Felt sheet

F Nail polish

G Needle

H Glue

I Chalk pen

J Scissors

K Awl

L Needle-nose pliers

Make It

Follow these steps to make your parallel circuit project:

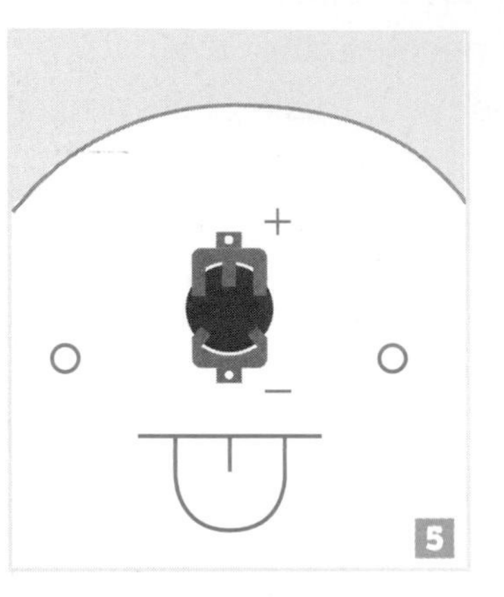

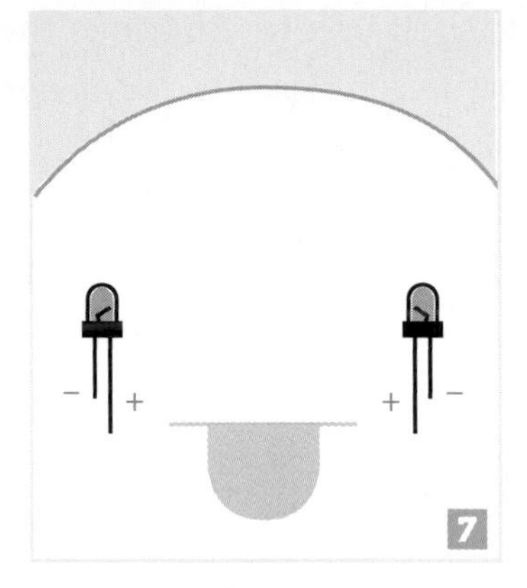

1. Draw the outline of your character on a sheet of felt using the chalk pen.

2. Cut out the outline of your character.

3. If you're making the ghost I'm demonstrating, use your scissors to cut a slit where you want the mouth located and insert the tongue so that it sticks out the front.

4. Glue the tongue to the back of the character so that it can still flap up and down in the front.

5. Using fabric glue, glue your battery holder on the back of your character. Make sure that the positive and negative sides of the holder are facing the right way.

6. Now, using an awl, poke four holes in the felt for the legs of your LEDs.

7. Using a felt pen, mark the holes for the positive and negative legs as in this illustration.

8. Place the legs of the LEDs into the holes, making sure that the long legs are in the positive holes and the short legs are in the negative holes.

9. After curling the legs of the LEDs on the back side of the character, connect the two positive legs of the LEDs with conductive thread.

10. Next, connect the conductive thread from the two positive legs to the positive side of the battery holder.

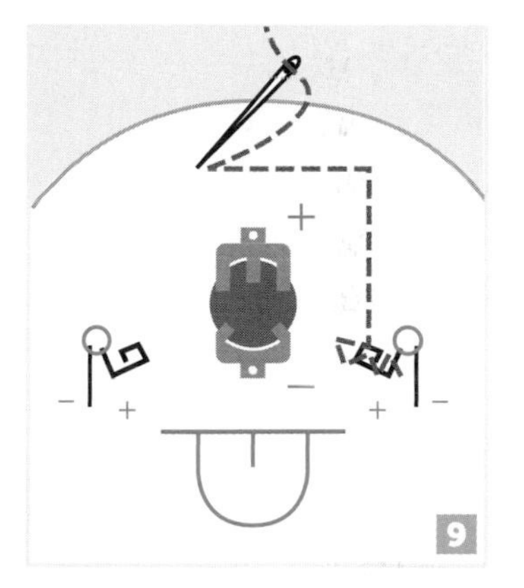

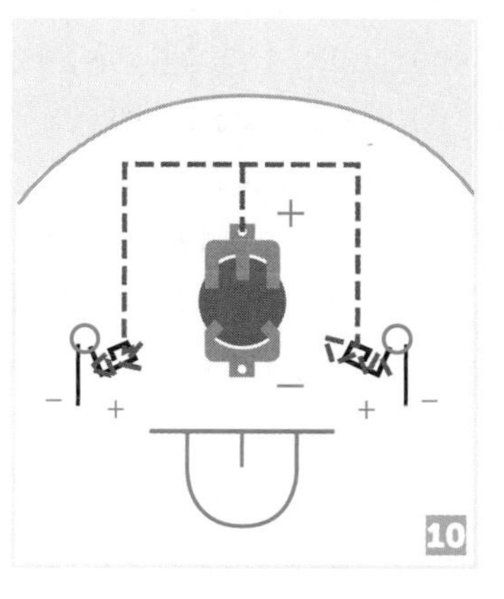

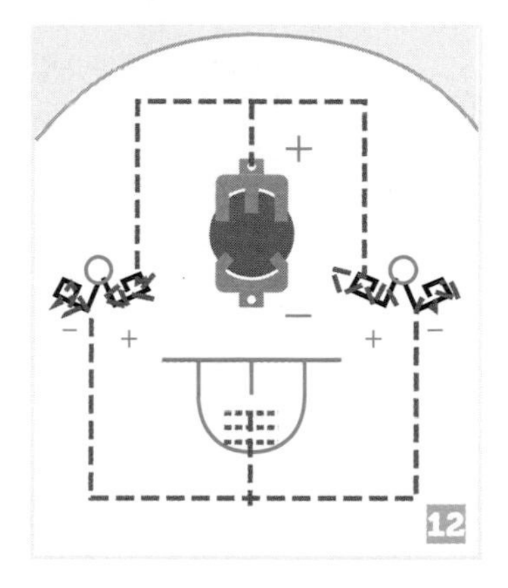

11. Connect the two negative legs of the LEDs with conductive thread.

12. Link this connection with one side of your switch. In my example, I sewed under the mouth of the ghost.

Sew back and forth a few times underneath the ghost's mouth. You want to make sure that there is enough conductive thread exposed for your switch to make a reliable connection.

13. Connect the negative side of the battery holder to the other side of your switch. In my example, I sewed it to the bottom of the ghost's tongue.

14. Again, make sure to sew back and forth a few times or in a circle so that there is enough conductive thread exposed for your switch to work well. See close-up of the two sides of the switch.

15. Put your battery in the battery holder and touch the two sides of the switch together. The LEDs should light up!

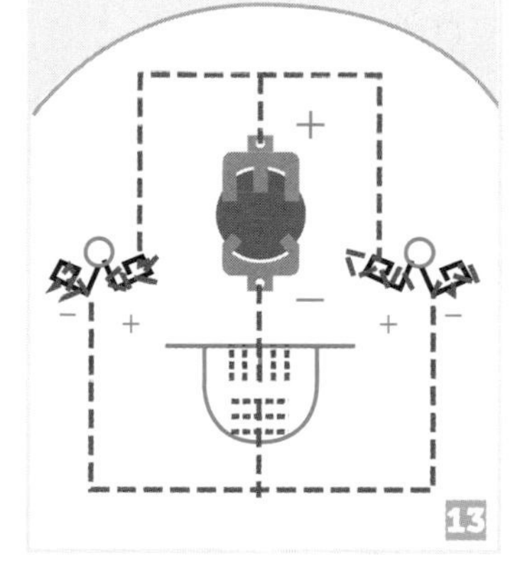

Figures 4.5 and 4.6 show the front and back of the finished ghost.

FIGURE 4.5: The finished ghost's silly smile

FIGURE 4.6: The back side of the ghost with completed circuit

By using a parallel circuit and using a switch creatively, as you did in the ghost project, you can make a wide range of characters and create some fun interactions.

For instance, you can try making up a story about your character so when you touch different parts of its body, lights turn on as shown in the Whoops! Robot pictured in Figure 4.7.

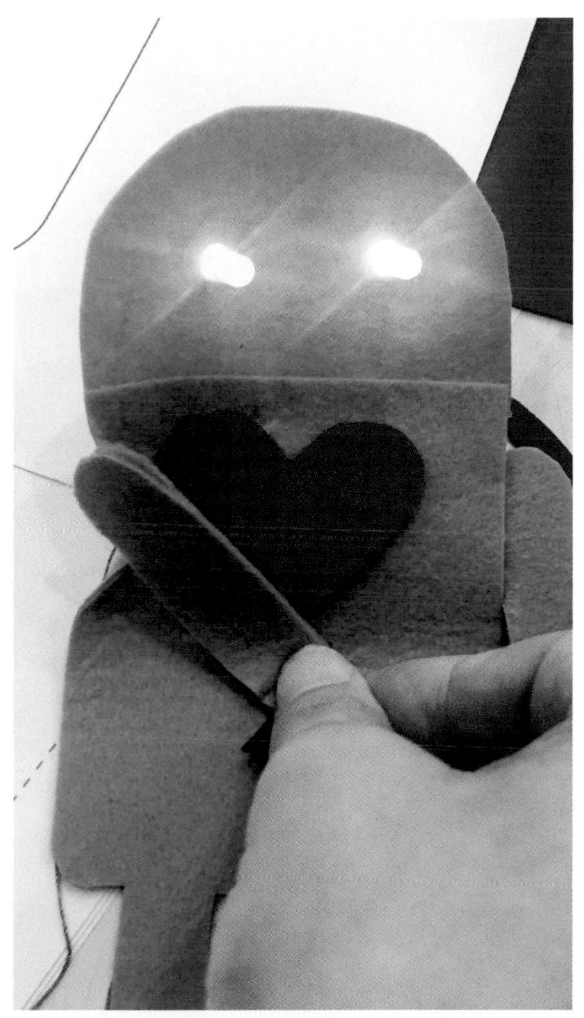

FIGURE 4.7: The Whoops! Robot

When I caught my daughter touching her belly button when she was a baby, her astonished reaction inspired me to come up with the idea for the Whoops! Robot. Touching his hand to his belly turns on lights in his eyes.

Figures 4.8 and 4.9 show a similar robot that was created in workshops using parallel circuits and switches.

Figures 4.10 and 4.11 show other fun projects—the Fish Lover Cat and the Pull Heart.

FIGURE 4.8: The Beeping Robot: When you use the robot's arm to press the button below the gauge on its chest, its eyes light up.

FIGURE 4.9: This is what the circuit on the back of the robot looks like.

FIGURE 4.10: This cat lights up her eyes when you place a fish inside her cat bowl.

FIGURE 4.11: This heart lights up when you pull the string.

Figure 4.12 shows a project called "A child that looks after me." Lights on her hair band light up when she holds her hands together. An elementary school student who participated in a workshop at the JUMP Festival 2011 in Korea made this project.

FIGURE 4.12: A project made by an elementary school student

The winking rabbit from Chapter 3 and this chapter's ghost both work on the same simple principle—using a switch to open and close an electric circuit. Using this same principle, you can create many fun projects.

In order to make the MP3 playing pillow shown in Figure 4.13, I first took apart an inexpensive MP3 player and found the connections that made up the control switches. I then connected conductive thread to those spots and extended the switches to conductive fabric. Then I used this idea to make a super large music pillow using a $13 MP3 player. This "Super iPod" project has received a lot of attention on the project website, Instructables.com. It even won an Editor's Selection Award! I still use the electronic sewing machine that I won in this contest. Figures 4.13 and 4.14

show the website where you can learn more about this project and see my friend James napping on the pillow.

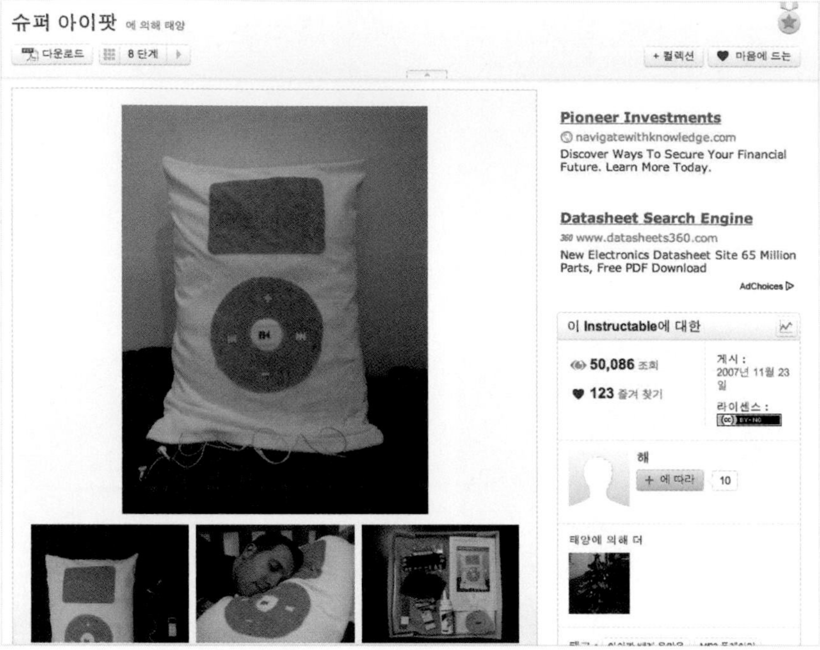

FIGURE 4.13: The instructions for making the Super iPod project are at www.instructables.com/id/Super-IPod-1/.

FIGURE 4.14: Taking a nap with the Super iPod pillow.

What's Next?

Now that you know how to add a second LED to make a series circuit, wouldn't it be nice to add even more LEDs? In the next chapter, you'll learn about voltage, current, and resistance as you make a project that incorporates up to ten LEDs!

Chapter
5

Letter Magnets

WORKING WITH LIGHT-EMITTING DIODES (LEDS)

In the modern world, we often see electronic signs, made of out of light-emitting diodes (LEDs), that are scrolling or flashing letters and words. These signs might tell us about anything from construction on the highway to the price of a sandwich in a deli. No matter what you want to say, to make letters with LEDs, you need to use a lot more LEDs and electrical connections than you have in your previous projects. In this chapter, you will learn about a few new concepts having to do with electric current, voltage, and resistance. In this project, you will make refrigerator magnets, like the ones in Figure 5.1, that are made up of many LEDs in the shapes of letters. You will use parallel circuits just like you did in the ghost project in Chapter 4 to make this magnetic alphabet.

FIGURE 5.1: Making your own light-up refrigerator magnets

In order to understand how to use many LEDs, it helps to know a bit more about the relationships between electric current, voltage, and resistance.

Electric Current

You can think of the electric circuit in a wire as being like water flowing through a pipe. In this case, the battery plays the role of a water storage tank, with gravity pushing electrons (water) through the wire (pipe). To continue this analogy, the quantity of water that passes by one point in the pipe over the course of one second is called the current. Likewise, electric current is a measurement of the number of electrons per second that are flowing past a given point in a circuit (see Figure 5.2). Electric current is measured in amperes, or amps for short.

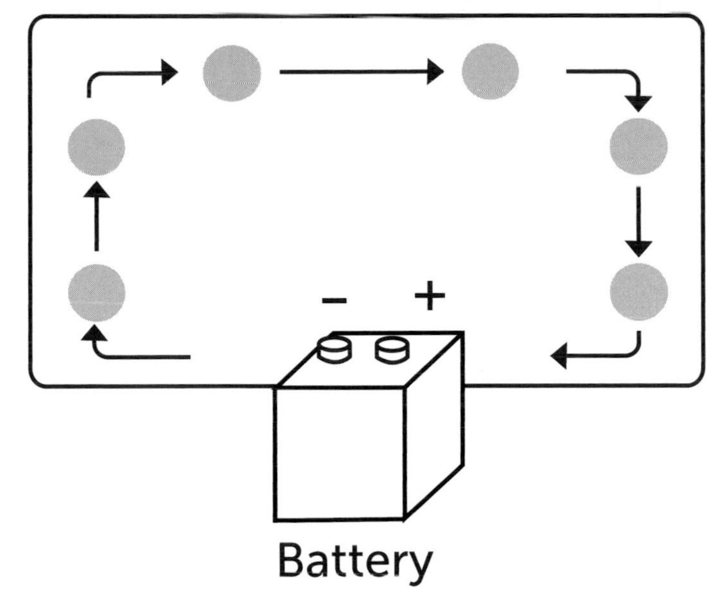

FIGURE 5.2: Electrons moving through a circuit

ELECTRON SPEED

A free electron flowing through a wire is not very fast; generally it moves less than three feet per second. So if this is the case, how can we use electricity flowing through very long wires to get near-instant communication or power transmission?

Free electrons in a wire are like the links on a loop of chain. If you pull on one link, they will all move almost at the same time. When an electric current starts in a wire, the free electrons in that wire all start moving almost simultaneously. So a signal sent from San Francisco to New York will cause the free electrons in the wire to move nearly at the same time.

Voltage

You can think of voltage as the pressure that pushes an electric current through a wire. Take a look at Figure 5.3. In our water analogy, the voltage is the pressure, due to earth's gravity, of the water in the storage tank pushing water through the pipe (1). If you double the amount of pressure by doubling the amount of water in the tank (2), this is like connecting two of the same batteries in a series—the voltage will double. But if you set up two storage tanks in parallel (3), the pressure remains the same, but the amount of water flowing through the pipe doubles. In this case the voltage is the same, but the current doubles. Voltage is measured in volts.

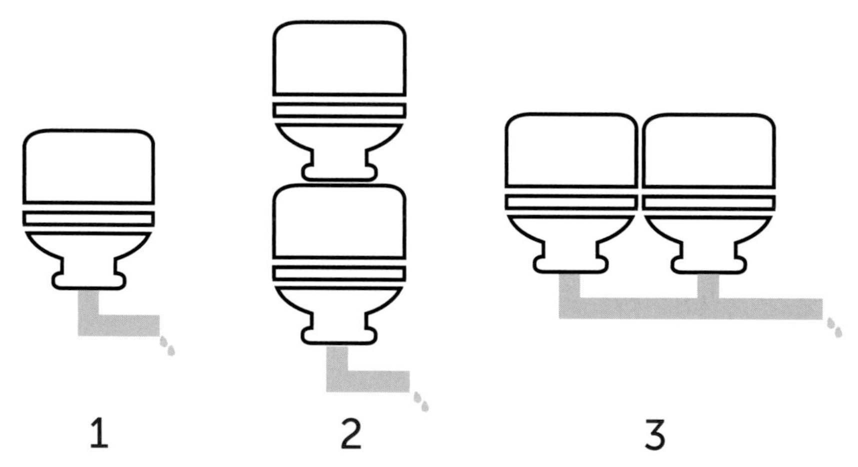

FIGURE 5.3: Water in a storage tank pushing water through a pipe

If we put two 3V (3-volt) batteries in a series and connect them to a light, the voltage across the light will be 6 volts (see Figure 5.4). That light will shine brightly and probably burn out quickly!

On the other hand, if we put two 3V batteries in parallel and connect them to the light, the voltage remains the same

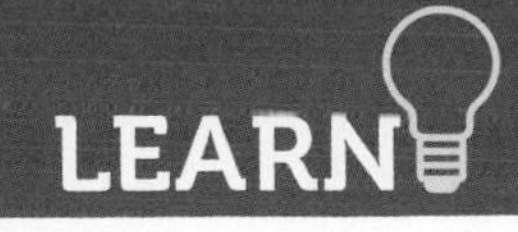

(see Figure 5.5). The light will shine at the same level as if there was one battery, but the batteries will last twice as long.

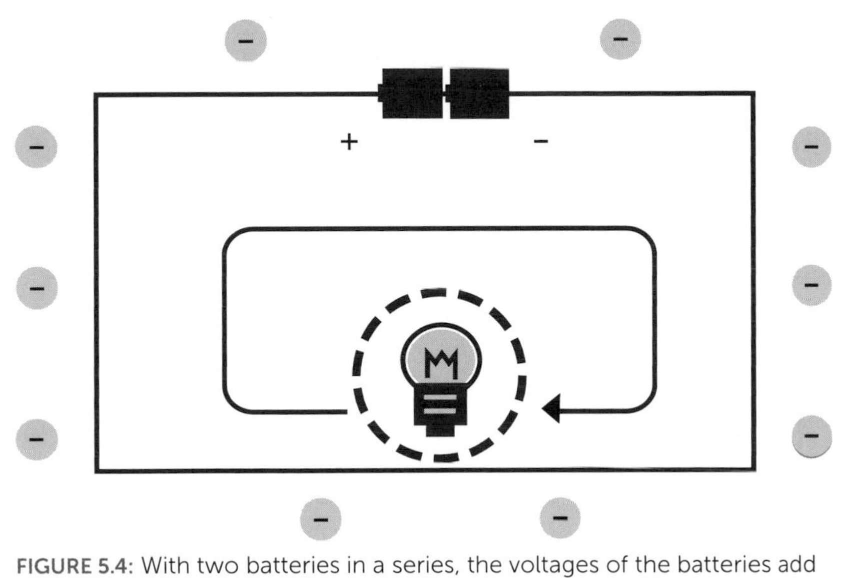

FIGURE 5.4: With two batteries in a series, the voltages of the batteries add up to the total voltage across the light.

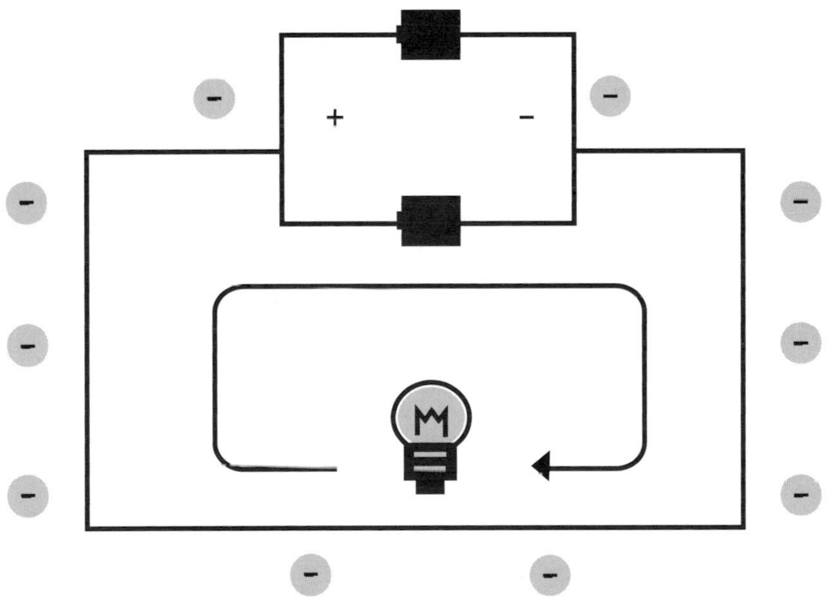

FIGURE 5.5: With two batteries in a parallel circuit, the voltage is the same as with one battery, but the charge will last longer.

Resistance

Any component in an electrical circuit will resist the flow of electrons in that circuit. The amount of resistance is measured in ohms. The symbol for ohm is the Ω. When you increase the amount of resistance in a circuit, the number of electrons flowing past any point decreases, so the current also decreases. In our water analogy, an increase in resistance could be caused by narrowing our pipe. A wide pipe allows a lot of water to flow through, so we say the current is high and the resistance is low. If we pump water through a narrower pipe at the same pressure, less water flows through that pipe over the course of a second, so we say the current is lower and the resistance is higher.

Relationship between Electric Current, Voltage, and Resistance

Since LEDs resist electric current, as you add more LEDs, the current decreases. This is not usually a problem when you are using two or three LEDs, but as you add many more, the current might get so low that your LEDs might start to dim or stop working altogether. There is a relationship between voltage, current, and resistance that allows you to figure out how much power you need to provide when you're working with many LEDs. This relationship is called Ohm's law.

This law tells you that if you increase the voltage in a circuit, you will increase the current. However, if you increase the resistance in a circuit, you will decrease the current. Your battery provides the voltage, and that remains basically constant (however, it does go down as the battery drains). Therefore, when you add more LEDs, the resistance in the circuit increases and the current decreases. But by adding a second 3V coin cell battery in parallel (see Figures 5.6 and 5.7), you increase the current and you can provide enough current for up to ten LEDs. You can use Ohm's law to calculate what kind of battery or batteries to use and how to regulate resistance in your circuit.

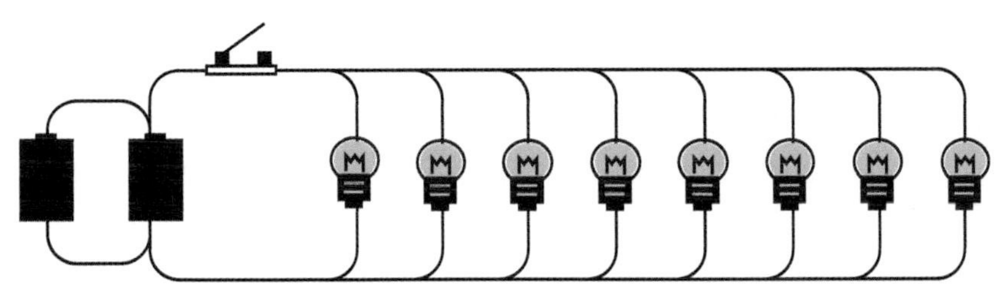

FIGURE 5.6: Adding a second battery in parallel doubles the current.

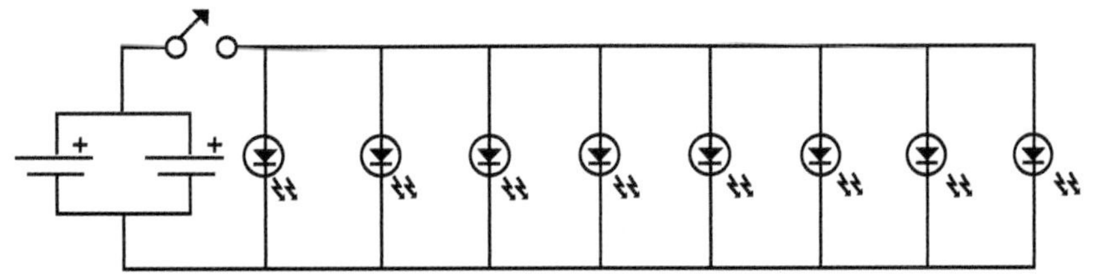

FIGURE 5.7: A circuit diagram of two batteries in parallel powering eight LEDs

Use Your Imagination

What are some of your ideas for using many LEDs in a circuit? How about making light-up letters from the alphabet? If you make several of these letters, then you can order and combine them to spell words! What letters do you want to make? What words do you want to be able to spell? In this project, we are going to put our letters on vegetables. Start by making a sketch of your project as in Figure 5.8. Make your sketch life-size and make sure that you have enough room for all of the components in your circuit.

Plan It Out

After you have sketched the project, add your circuit diagram to the sketch. Since you will form letters by connecting many LEDs in parallel for this project, it's really important to separate the positive and negative sides of your circuit. You

FIGURE 5.8: A sketch of a pepper onto which we are going to sew a circuit with many LEDs

are going to connect all of the positive legs of the LEDs to the positive sides of the batteries and all the negative legs of the LEDs to the negative sides of the batteries. Make sure to use no more than ten LEDs and connect two batteries to your circuit in parallel as illustrated in Figure 5.9.

Preparation

After you are done sketching and carefully planning the circuit, it is time to collect your materials, cut out your fabric, and sew your circuit. Since this circuit includes many LEDs, you have to be extra careful that you don't cross the threads of the positive and negative sides of the circuit. Take your time. Try to make each stitch as neat as possible. When you are looping the thread around the curled legs of your LEDs, loop it a few times to get a good connection, and then move on to the next LED.

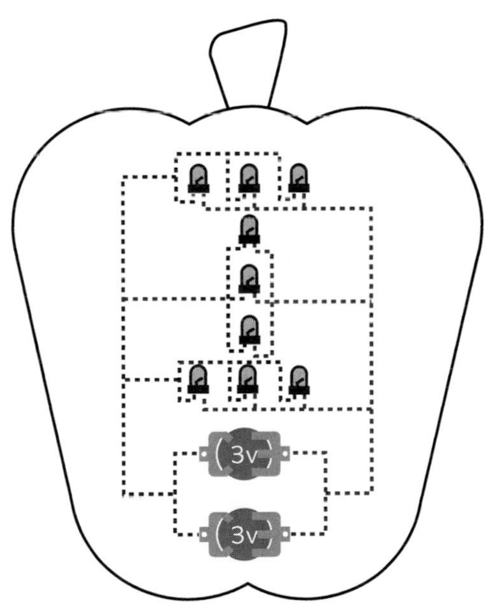

FIGURE 5.9: A sketch of the pepper that includes the circuit diagram. The positive side of the circuit is red. The negative side of the circuit is black.

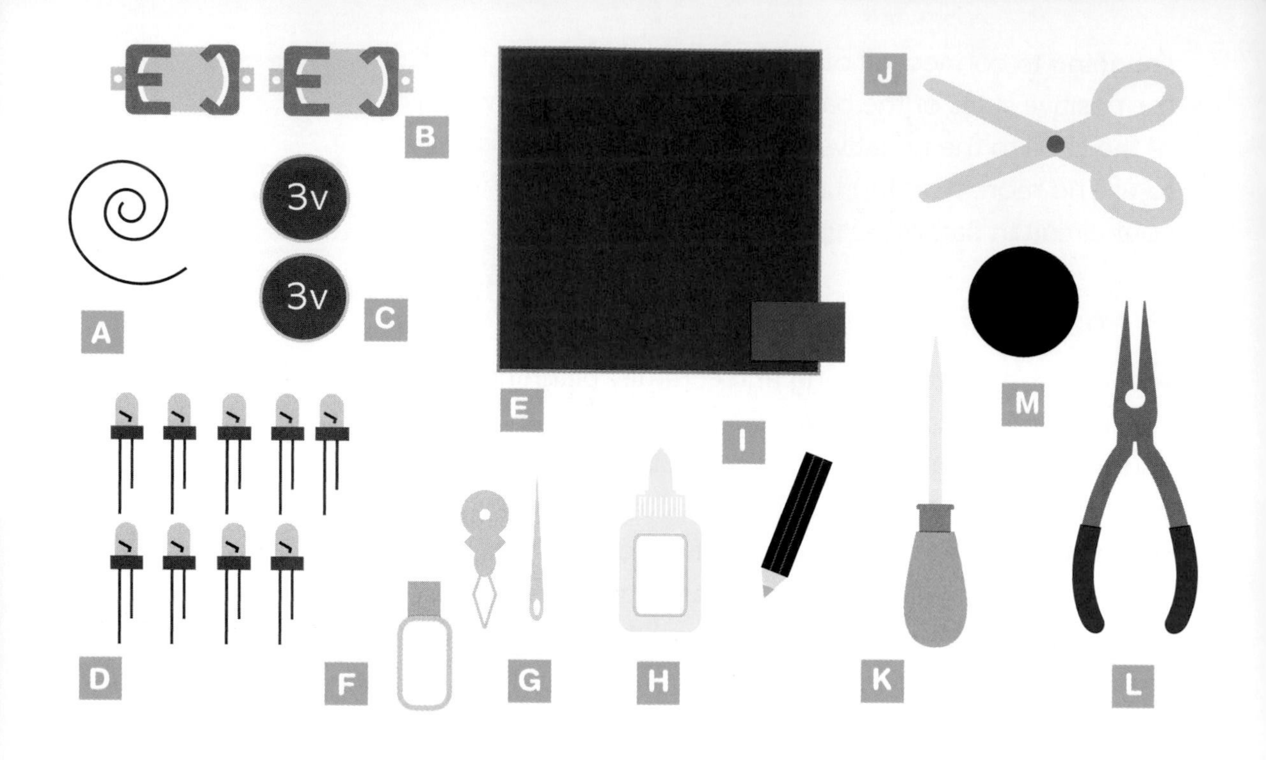

LIST OF SUPPLIES:

- A Conductive thread
- B Battery holders
- C 3V coin batteries
- D 9 LEDs
- E Felt sheet
- F Nail polish
- G Needle and threader
- H Glue
- I Chalk pen
- J Scissors
- K Awl or other sharp tool for poking small holes in fabric
- L Needle-nose pliers
- M Magnet

Make It

Follow these steps to make a light-up letter project with several LEDs:

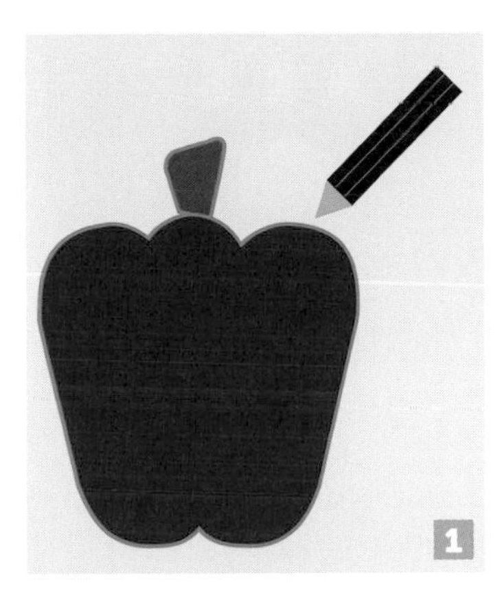

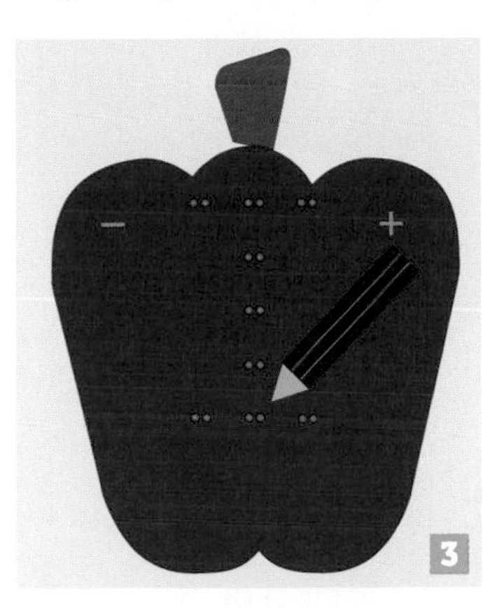

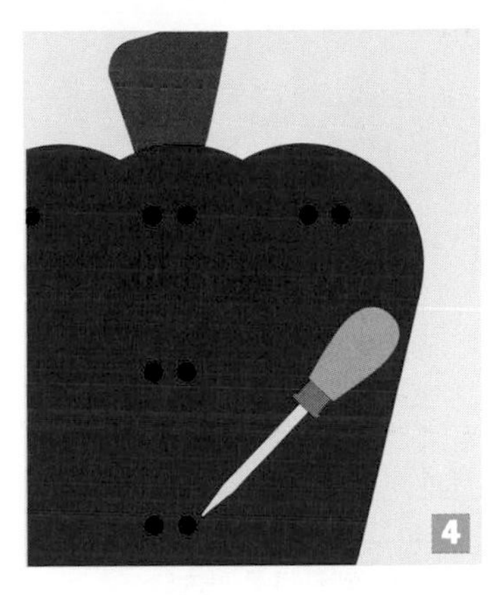

1. Use a chalk pen to sketch the outside of your shapes on felt sheets.

2. Cut out the shapes and use fabric glue to paste any pieces together.

3. Use a chalk pen to lightly draw dots in the pattern of your letter for the legs of the LEDs. Mark the positive and negative sides of your project to remind yourself to keep the positive holes on one side of the circuit and the negative holes on the other side of the circuit. Make sure to leave room on the felt for the batteries.

4. Using an awl, or other sharp tool, carefully poke holes for the legs of the LEDs.

5. Place the legs of your LEDs into the holes from the front so that the legs come through to the back of the felt. Take extra care to make sure that the positive (long) legs and the negative (short) legs go into the correct holes.

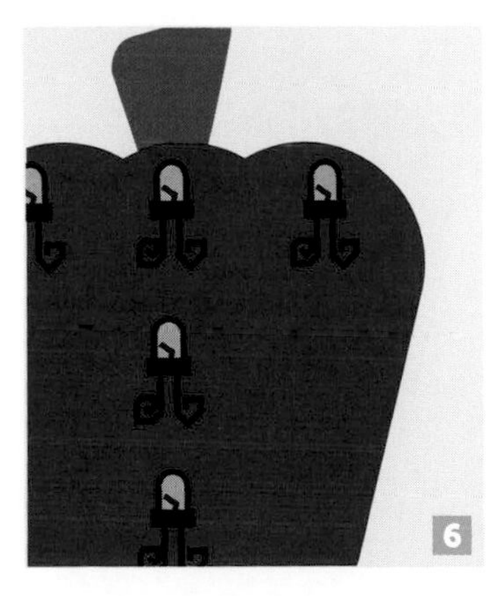

6. Use needle-nose pliers to curl the legs of the LEDs on the back of the felt sheet as you have done for the previous projects. Take extra care to make sure that the positive and negative legs of the LEDs don't touch each other. *Note that in these images, the lights of the LEDs are shown on the back for clarity.*

7. Place the battery holders on the felt sheet. Double check the orientation of the battery holders. Make sure the positive and negative sides are facing in the same direction as the LEDs.

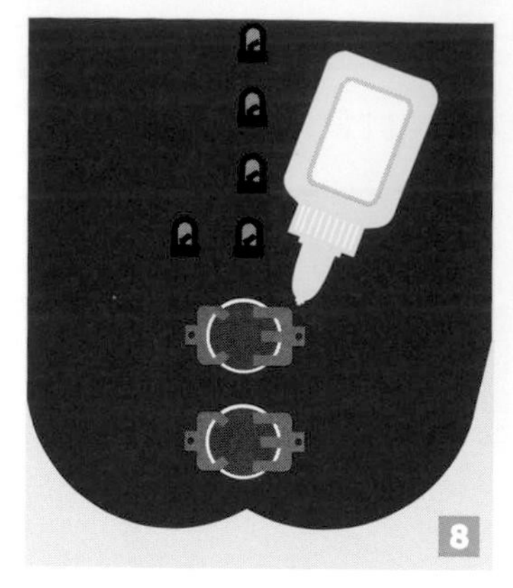

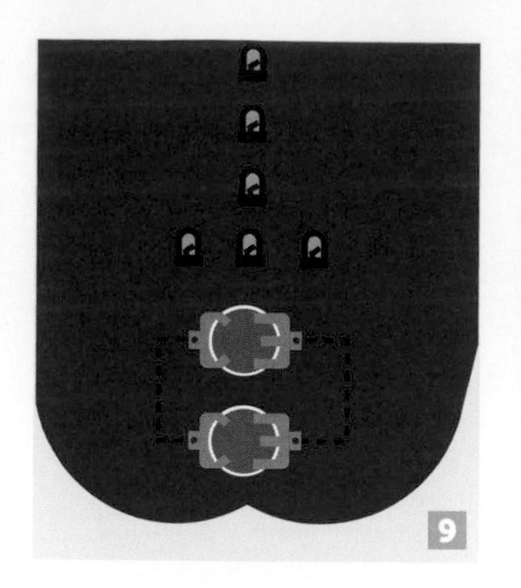

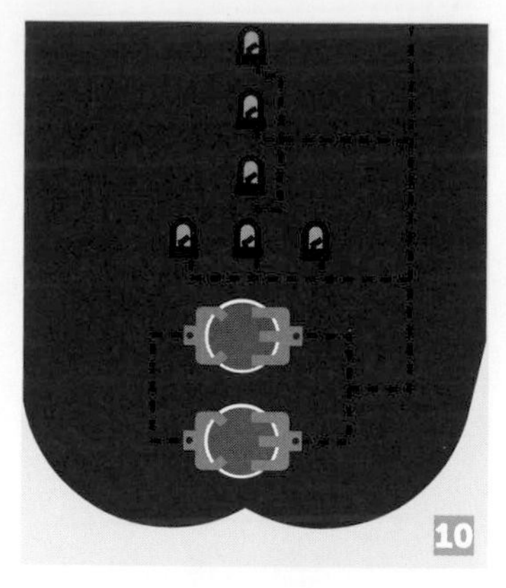

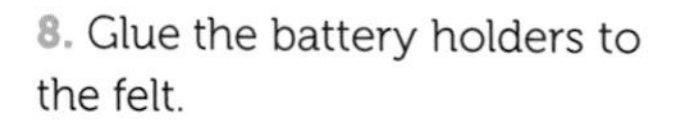

8. Glue the battery holders to the felt.

9. Now you can start sewing your circuit. Use conductive thread to connect the positive sides of your battery holders together. Then use another piece of conductive thread to connect the negative sides of your battery holders together.

10. Next, use the conductive thread to connect all of the positive legs of the LEDs together.

You can use the same piece of thread to connect all the LEDs; just loop it through each curled leg twice to get a good electrical connection, and then tie the thread to the positive connection of the battery holders.

11. Repeat the last step with the negative side of the circuit.

12. Now insert batteries into the holders to test your circuit.

Once you are sure that your circuit is working, cover the circuit with a sheet of felt and glue it down. This is to prevent a short circuit if you attach your letter to a metal surface. Make sure not to cover the battery holders so that you can still change the batteries once they run out of charge.

13. Finally, glue a magnet to the back side of the felt.

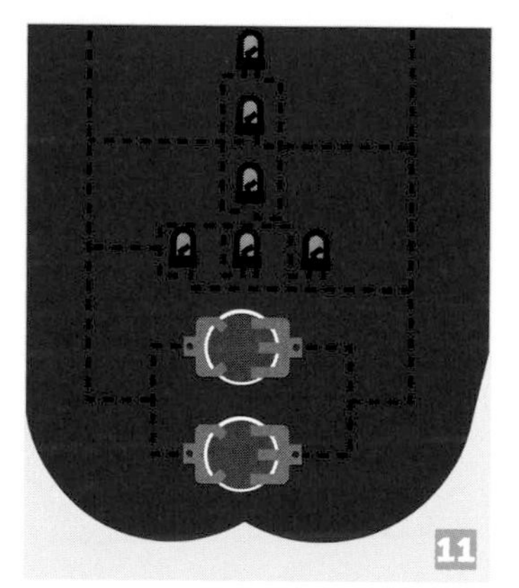

Figure 5.10 shows what the back side of the finished magnet looks like.

FIGURE 5.10: The back of a turnip with a felt sheet covering the circuit and a magnet attached

Now you're all done (see Figure 5.11)! Place your magnetic LED letters on your refrigerator or any other metal surface.

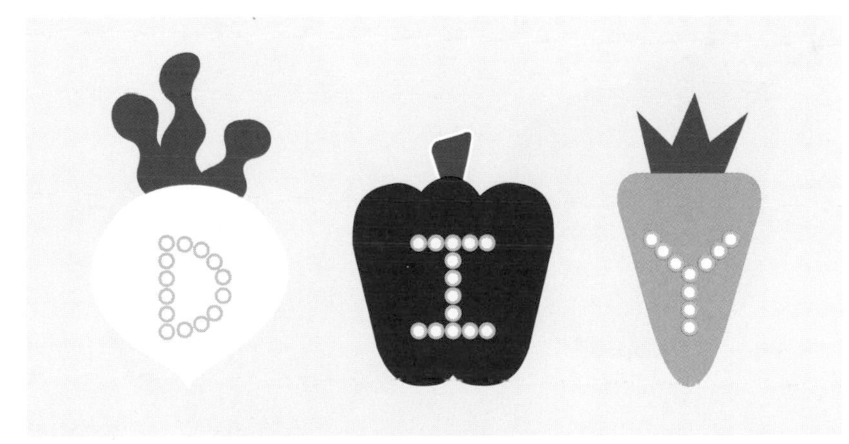

FIGURE 5.11: Completed LED letter magnets

Proudly show off your project. Make a word with your letters and take a picture or share with your friends and family.

Now it's time to explore some other works that use electric current, voltage, and resistance to light up many LEDs.

Figure 5.12 shows a ladybug with firefly aspirations.

FIGURE 5.12: A light-up ladybug (front and back)

Which Legs Turn On the Light?

This monster octopus with many eyes (see Figure 5.13) was made by an elementary school student. Since the child wanted to add some sort of playful component to the project, only two of the legs turn on the monster's eyes when they are connected to each other.

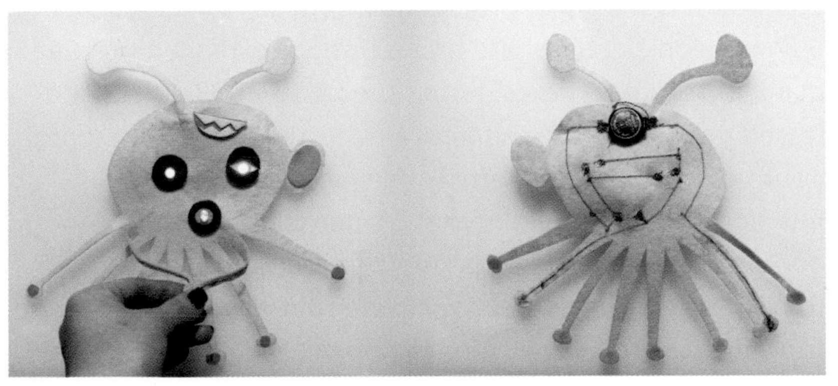

FIGURE 5.13: Playful octopus (front and back)

Flower Decorated Headband

This headband was the result of a collaboration between an adult and a child. Some young girls like headbands that attract attention. The LEDs serve that purpose. How might you add a switch to this project so that you can turn it on and off when you need to?

The Secret Tree

LEDs are used in a lot of technological artwork because they are inexpensive and they have a nice visual effect. But as you learned, when you are using many LEDs, it can be difficult to maintain the optimal relationships between voltage, current, and resistance.

I made the Secret Tree project that appears in Figures 5.14–5.17 so people could share their secrets. I exhibited this tree at a Maker Faire in Texas during the fall of 2008. If you confessed your secrets to the tree, a microphone would record your voice and then scramble the words. You could also interact with the tree to hear the scrambled voices of past confessions. This made it possible to recognize voices, but the content of the confessions were unrecognizable. I added moonlit flowers to the tree that brightened and dimmed according to the volume of the voices of the people recording their confessions.

FIGURE 5.14: A moonlit flower on the Secret Tree

FIGURE 5.15: Testing the circuit for the Secret Tree

FIGURE 5.16: Listening to scrambled confessions from the Secret Tree

FIGURE 5.17: Confessing secrets to the Secret Tree

I started prototyping this project with parchment paper and LEDs that I connected in parallel.

As you can imagine given the size of the final tree and the many moonlit flowers, I had to use Ohm's law to determine the proper power source that I would need to power the hundreds of LEDs in the project.

If you understand how to use Ohm's law, you can apply it to many types of projects. Figures 5.18 and 5.19 show a recycled shopping bag that I turned into a miniature version of the secret tree.

FIGURE 5.18: Recycling materials for projects

FIGURE 5.19: A miniature version of the Secret Tree made from recycled materials

What's Next?

So far you have been making circuits with LEDs; it's really fun to make things that light up. But in the next chapter, you'll do something even more fun; you will learn how to use a vibrating motor to make a project that moves!

Chapter 6

Purring Elephant Pillow

ANIMATING WITH MOTORS

Congratulations! So far you have made several projects that light up using light-emitting diodes (LEDs). The last one even used up to ten LEDs per letter! That is a lot of sewing. Now it is time to try something a bit different. In this project, you are going to learn how to use a motor to make a project that moves.

When you think about motors, you probably think of things that spin. Toys that move, a kitchen blender, and clocks all contain motors that turn. But did you know that cell phones use motors too? When a cell phone buzzes, or vibrates, you are feeling the movement of a special type of motor called a vibrating motor. In this project, you are going to be using this type of motor to make a massaging elephant pillow (see Figure 6.1).

Electricity and magnetism are closely related. If you send an electric current through a wire, a magnetic field will appear around that wire. Conversely, if you spin a magnet around a wire, you create an electric current running through that wire. This relationship between electricity and magnetism is what makes a motor work.

A motor contains a magnet and a length of wire that is coiled up into many loops. When you run electricity through

FIGURE 6.1: This purring elephant has a vibrating motor sewn into the back of his head.

the loops of wire, you create a magnetic field that moves a magnet attached to the axle of the motor. This makes the axle of the motor spin.

Vibrating Motors

A vibrating motor is just a regular motor with an unbalanced weight on the rotating axle. As the motor inside spins, the unbalanced weight causes the motor to shake back and forth. This is what causes your cell phone to buzz when it vibrates to indicate you have a message. Figure 6.2 shows a circuit diagram that includes a battery, switch, and motor.

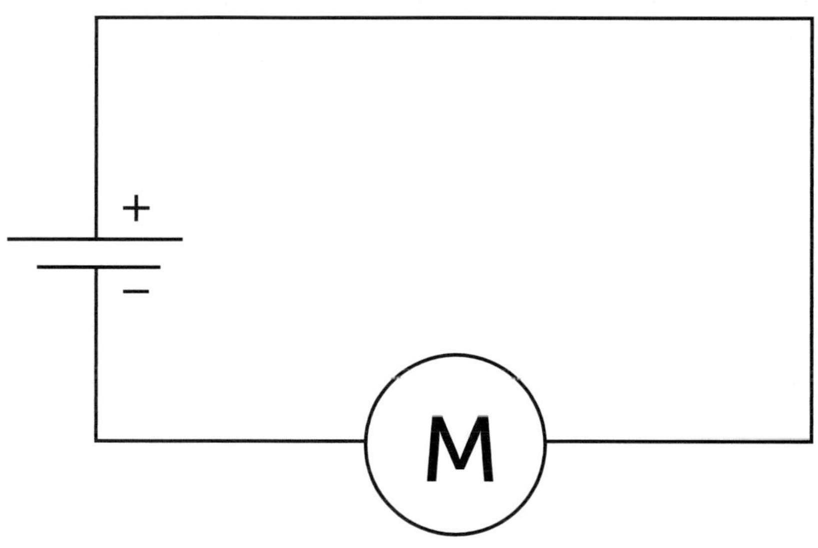

FIGURE 6.2: Circuit diagram of a vibrating motor with a switch

Use Your Imagination

Did you know that, like cats, elephants purr? That's right. The largest land animals can communicate over long distances by making a low purring sound and vibrating the earth! In this project you are going to use a vibrating motor to make a project that purrs and shakes. In the following pages, I'll describe how to make an elephant that buzzes and vibrates (see Figure 6.3). What design do you want to make? Sketch out your design on paper.

Plan It Out

This is a very simple circuit with a battery and a motor. You are going to put the motor inside the project (Figure 6.4), but make sure to keep the battery holder on the outside (see Figure 6.5) so that you can take the battery out and change

FIGURE 6.3: An illustration of the moving elephant

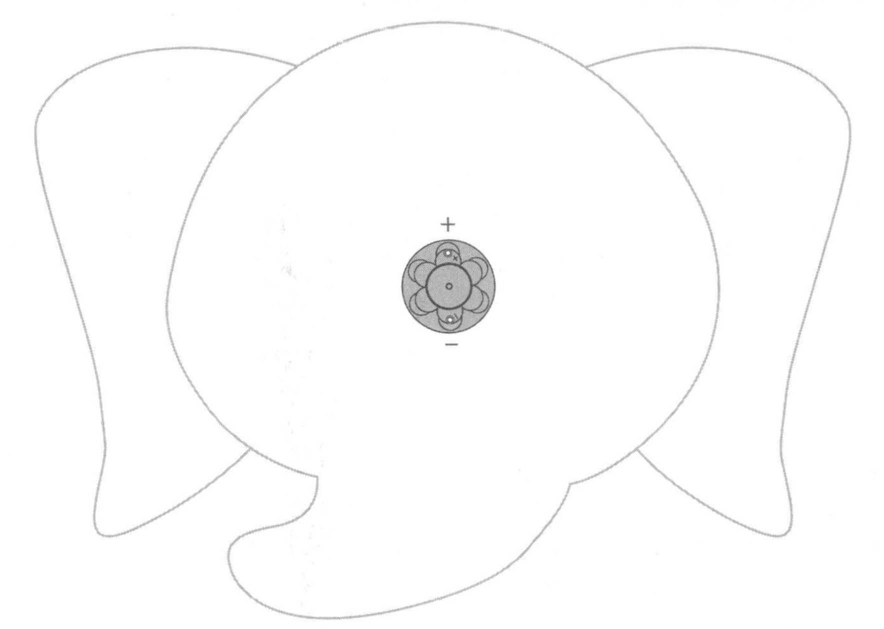

FIGURE 6.4: The inside of the purring elephant sketch with a drawing of the motor

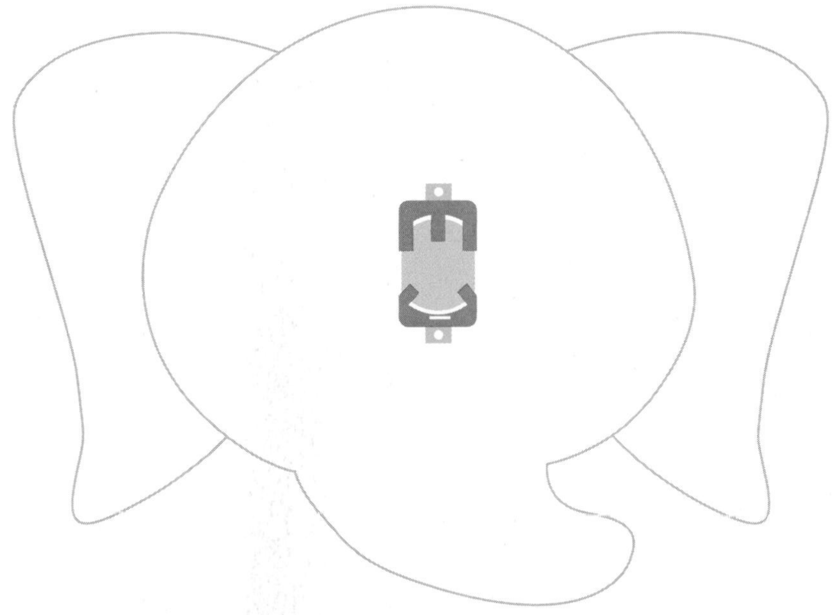

FIGURE 6.5: The back of the purring elephant sketch with a drawing of the battery holder

it when you need to. Sketch out where you are going to place the battery holder and the motor. Make sure to mark the positive and negative sides of your circuit.

There are several types of vibrating motors. Many of them have very small wires that are difficult to attach to conductive thread. I recommend using the LilyPad Vibe Board because it is designed for sewing circuits and it is easy to use. You can order this from several places online. Check the book's website for a list of recommended suppliers: www.techdiy.org.

Preparation

Once you have the outline and circuit for your project sketched out, you should gather your tools and materials. Make sure to read all of the steps in the "Make It" section before you start so that you understand where the pieces are going to go before you start gluing and sewing the project together.

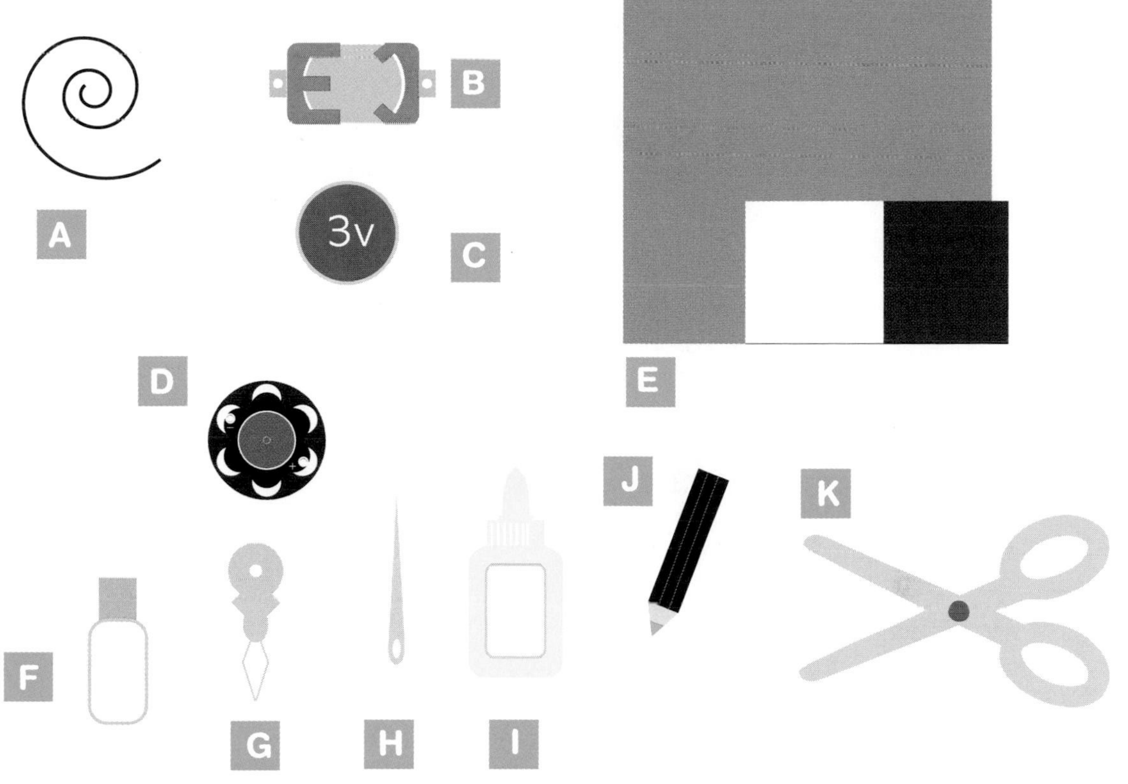

LIST OF SUPPLIES:

A Conductive thread

B Battery holder

C 3V coin cell battery

D LilyPad Vibe Board

E Felt

F Clear nail polish

G Needle threader

H Sewing needle

I Fabric glue

J Chalk pen

K Scissors

Make It

Follow these steps to make your vibrating motor circuit project:

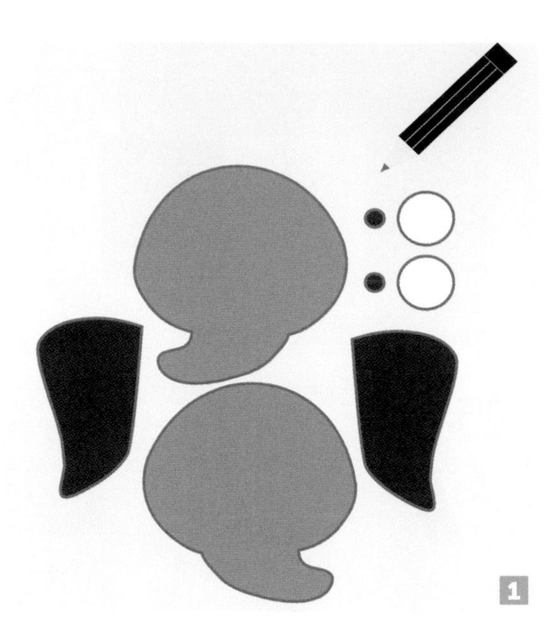

1. Based on your paper sketch, use a chalk pen to sketch the pieces of your character onto colored felt. Make sure to trace two copies of the main part that will house the circuit. One of them will be for covering the front of your circuit and one will be for the back.

2. Cut out all of these pieces.

3. Using fabric glue, glue all of the pieces together except the back piece. That is where you are going to sew your circuit.

4. Flip over the back piece of felt and glue the battery holder to the outside of the back sheet. Make sure that the positive side of the battery holder is facing toward the top of your project.

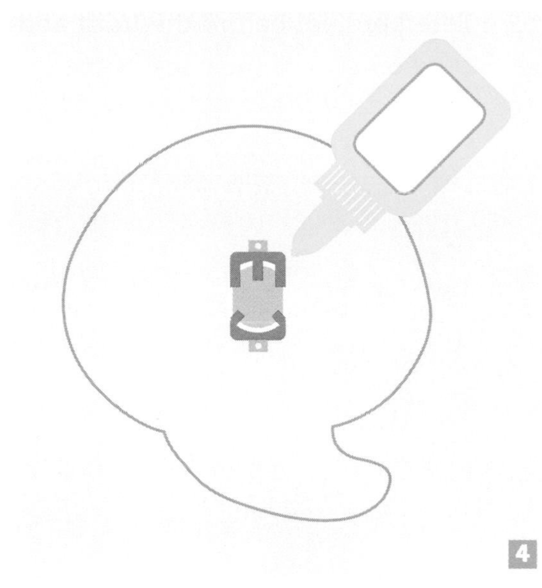

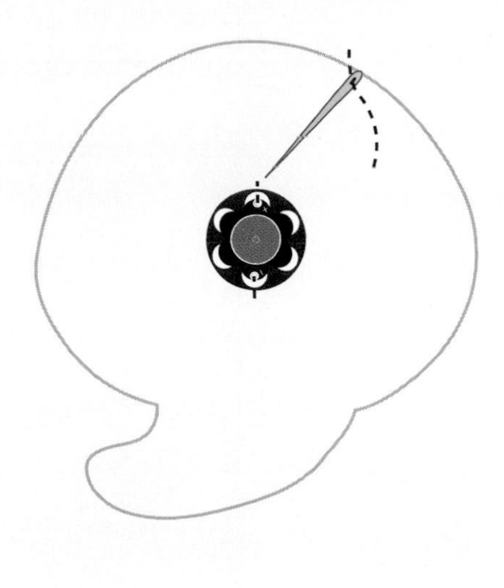

5. Flip the back piece of felt over so that what will be the inside of the back piece is facing you. Place the LilyPad Vibe Board here, on the opposite side of the felt piece from the battery holder. Make sure that the positive side of the Vibe Board, marked with a plus sign (+), is facing toward the top of the project.

6. Using conductive thread, sew through the piece. Attach the positive side of the Vibe Board to the positive side of the battery holder. Do the same with the negative side. Remember that the battery holder will be on the outside of the back of your project, and the motor will be on the inside of the project.

7. Test the circuit by putting a battery in the battery holder. The motor should start buzzing and vibrating. Once you are sure that it is working, take out the battery.

8. Using fabric glue, glue the front of the project to the back piece of the project.

Figure 6.6 shows the completed project before the front and back are glued together. Figure 6.7 shows the back of the completed project.

FIGURE 6.6: The front of the elephant with the back piece and Vibe Board motor. This is before the pieces are glued together.

FIGURE 6.7: The back of the completed project with the battery holder exposed

Now that you've made the purrfect vibrating project, it's time to see what else can be made with similar motors.

This massaging headband was made by a young student at the The Makery in New York City. During a soft circuitry workshop, she used a battery to power two vibrating motors. She added a pink flower that you could remove with two snaps. This flower was decorative but it also served as a switch so that she could turn it on or off.

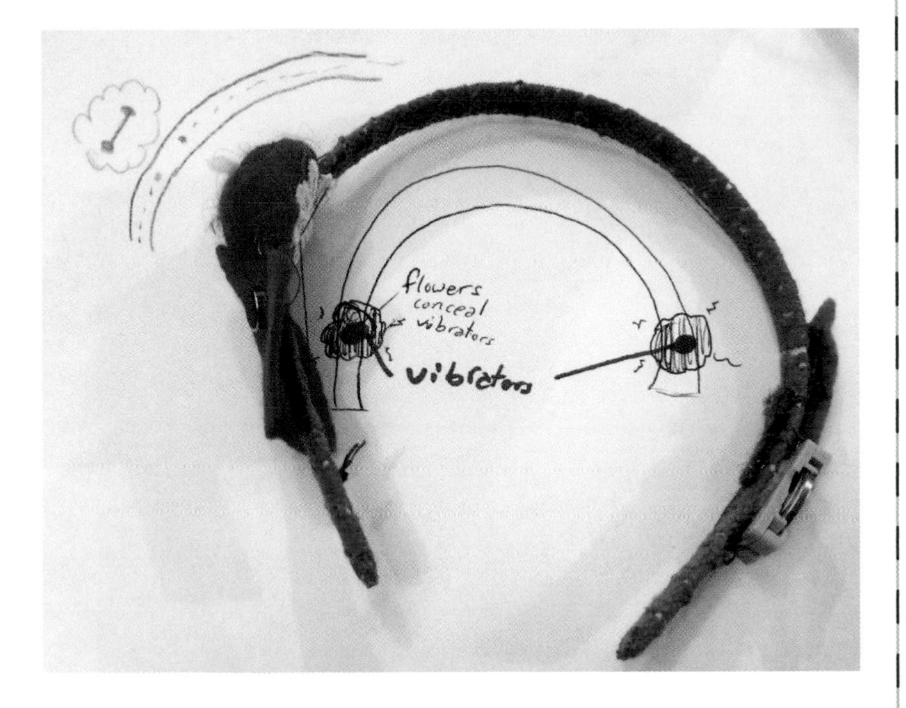

Figures 6.8 and 6.9 show another example that you might try. These images are of a neck massager that I made. A vibrating motor in the middle of the project gives you a massage when you squeeze a button in the ear.

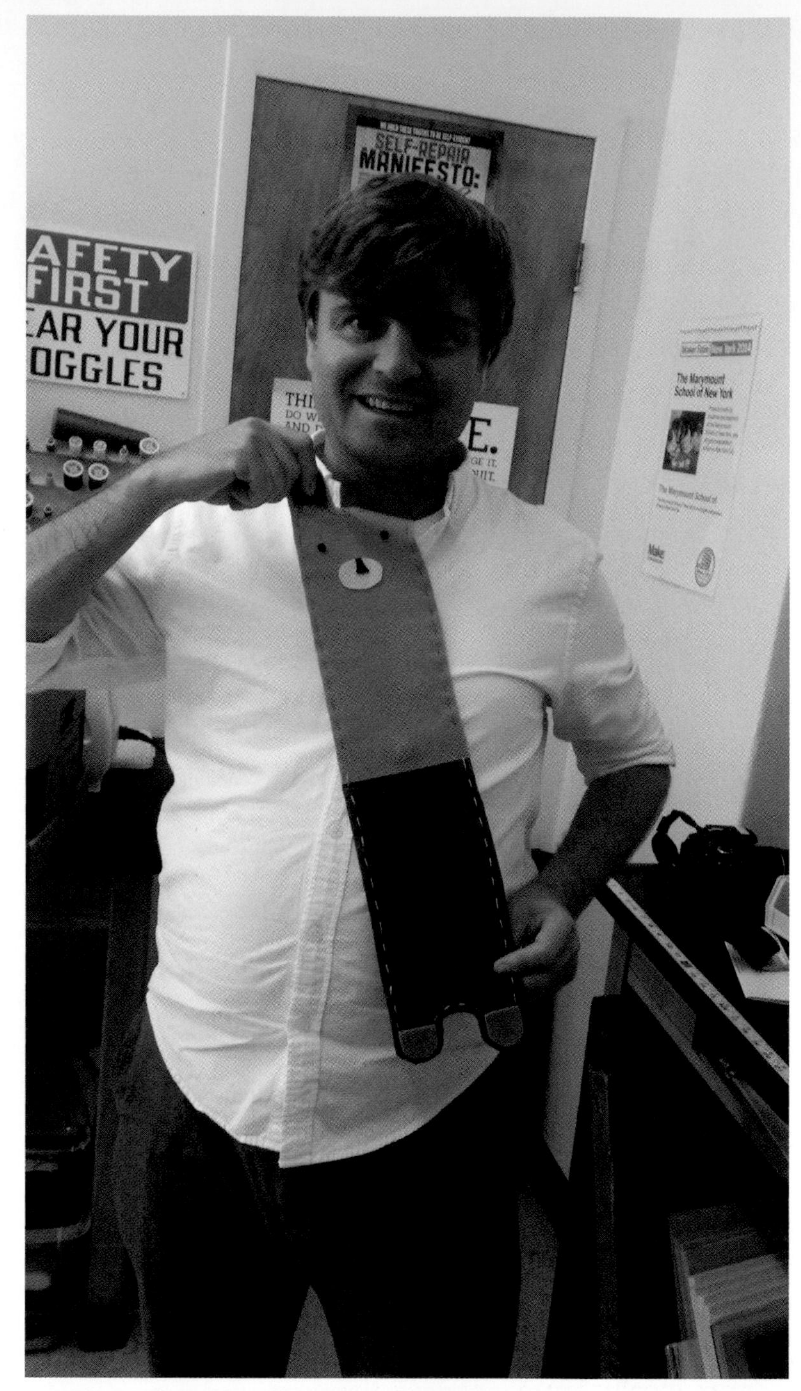

FIGURE 6.8: Showing off the neck massager

FIGURE 6.9: The neck massager being put to good use

What's Next?

In this chapter, you took a bit of a break from sewing with LEDs. You learned about motors and how to use them. In the next chapter, you will go back to using LEDs. You are going to learn how to use a light sensor to make a project that automatically lights up in the dark!

Chapter

7

Nightlight Cat Bracelet

WORKING WITH LIGHT SENSORS

Sensors are electronic components that detect different traits about the environment around them. In fact, you already used sensors in the projects in Chapters 3 and 4. A switch or button is a type of sensor that senses whether it is being pressed. There are also sensors that sense things like temperature, humidity, noise volume, and much, much more. In the modern world, we are surrounded by many types of sensors that are used in our electronic devices. Smartphones are packed with sensors that measure tilt, sound, touch, and more.

In this project, you are going to make a cat bracelet that has a light sensor. This is the same type of sensor that is in a nightlight. This cat bracelet automatically turns on the lights in its eyes when it senses darkness.

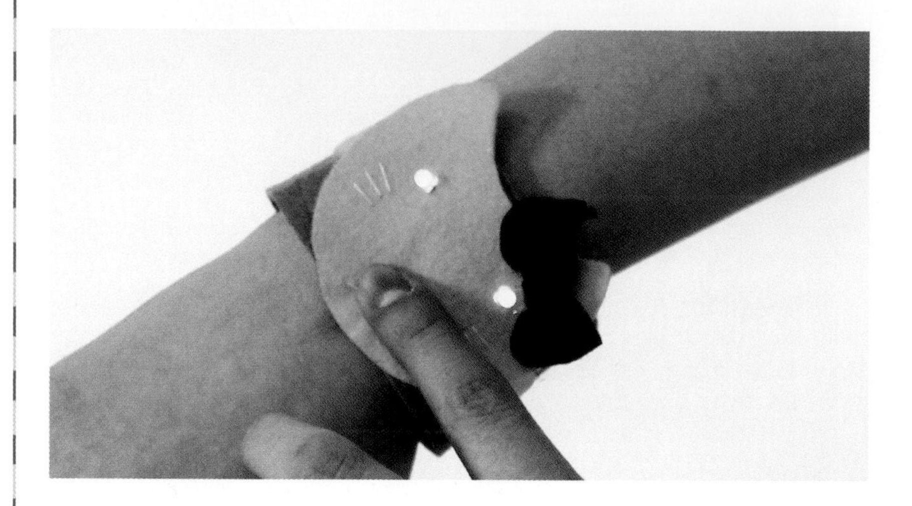

Let's learn about two new components: transistors and light sensors. Once you're familiar with these, you can make a circuit that uses them to turn on two light-emitting diodes (LEDs) when it gets dark.

Transistors

Transistors are the most common electronic components found in today's digital devices. Transistors have three parts: a base, a collector, and an emitter (see Figure 7.1). You can think of the base of a transistor as a switch that controls the flow of electricity from the collector to the emitter. If there is no electric current applied to the base, no current flows from the collector to the emitter. But by applying a small electric current to the base of a transistor, you can allow a larger current to flow from the collector to the emitter; the

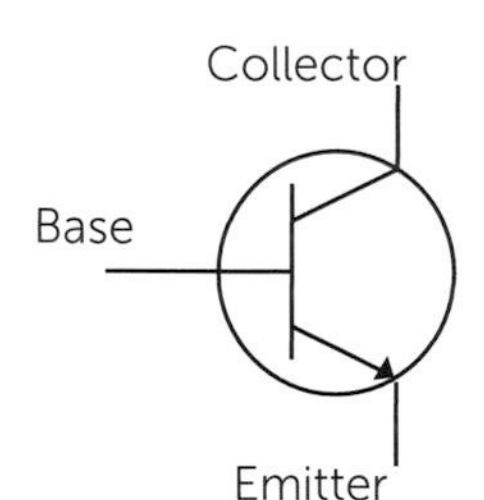

FIGURE 7.1: Electrical symbol for a transistor

more current you apply to the base, the more current flows through the collector and emitter, as shown in Figure 7.2.

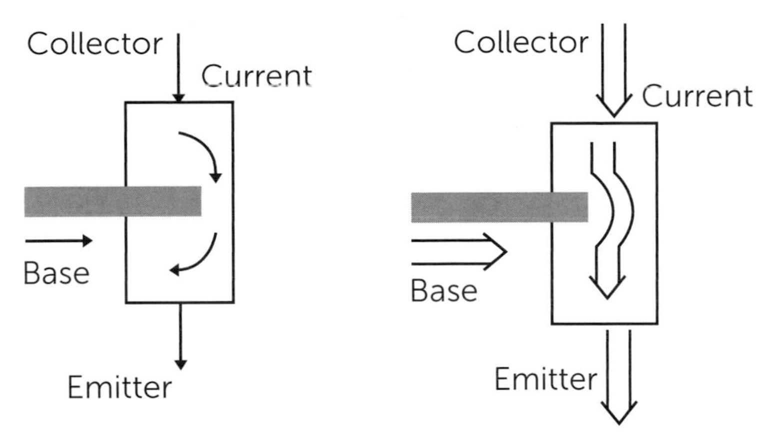

FIGURE 7.2: Applying a greater current to the base allows more current to pass through the collector to the emitter.

Electricity will only flow from the collector through the transistor and out the emitter if a current is applied to the base of the transistor. That is why you can think of a transistor as a switch that is opened and closed by applying electricity to the base. Figure 7.3 depicts an illustration of a 2N3904 transistor. Notice that the front side of the transistor is flat. It is important to know which side of a transistor is the front so that you put it into your circuit properly.

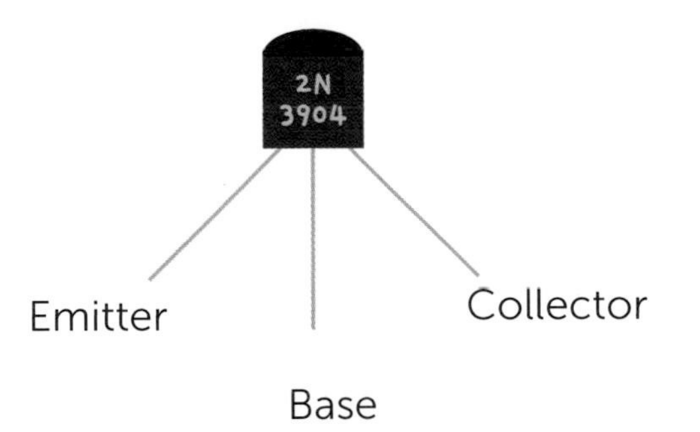

FIGURE 7.3: These are the locations of the base, collector, and emitter on a 2N3904 transistor.

Light Sensors

Sensors are like the human sensory organs of the electronics world. Light sensors, also called *photocells*, are like inexpensive eyes. This type of sensor uses a material called cadmium sulfide to measure how much light is shining on the sensor (see Figure 7.4). When it is dark out, or if the sensor is covered by a shadow, the electrical resistance through the cadmium sulfide increases; as a result, it does not conduct electricity very well. If it is light out, or if you shine a light on the sensor, the resistance through the sensor decreases. Basically, a light sensor is more conductive when it is exposed to light and less conductive when it is in the dark. Unlike with LEDs or transistors, the direction in which you orient the light sensor in your circuit does not matter. It will work the same either way.

Light sensors are used in electronic devices to find the level of light in an environment. These sensors are often used in

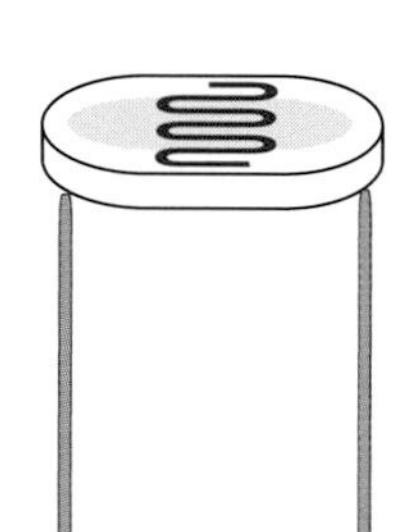

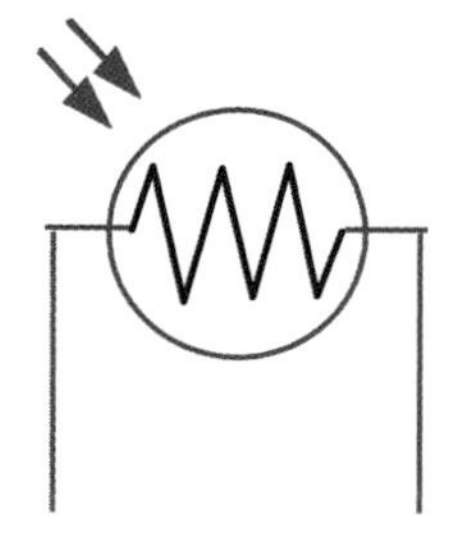

FIGURE 7.4: A photocell and its electrical symbol. The squiggly red line is the cadmium sulfide.

nightlights because they can save electricity by only turning on when it is dark. These types of light sensors are not very precise, but they are extremely inexpensive, and they are perfect for this nightlight cat circuit.

The Circuit

Figure 7.5 shows a circuit that turns on two LEDs when the light sensor detects a low level of light. This circuit might look a bit confusing. But if we remember how the transistor works and then follow the path of electricity when it is light and when it is dark, we can learn how this circuit works.

When light shines on the light sensor, it is more conductive, and electricity can flow through it easily. Since electricity wants to flow through the path of least resistance, the electricity will pass through the circuit colored in red in

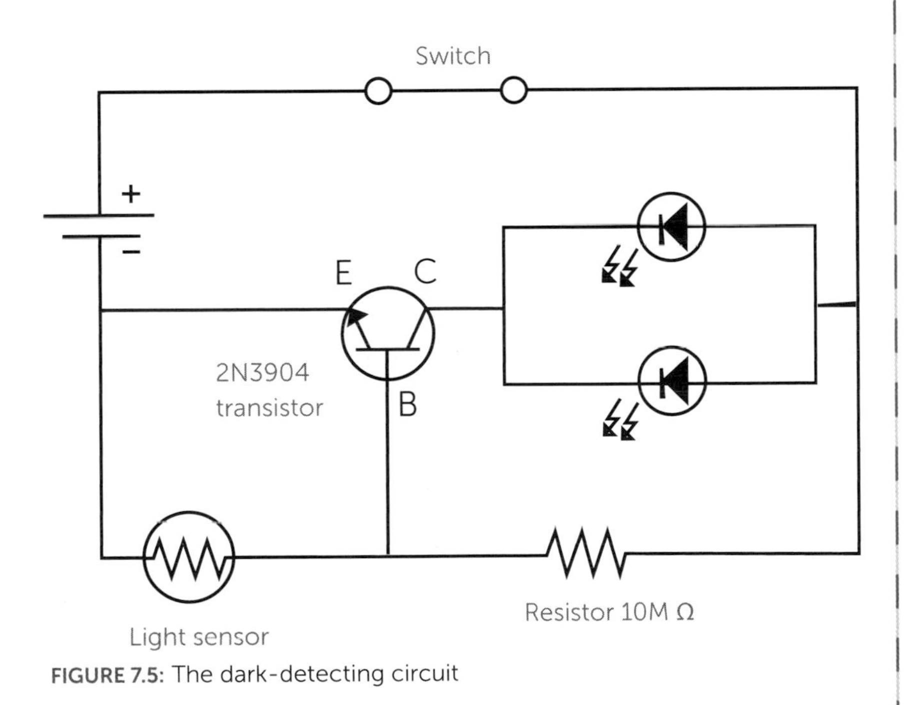

FIGURE 7.5: The dark-detecting circuit

Figure 7.6. It goes through the 10M Ω resistor and the photocell. 10M Ω is short for 10 megaohms, or 1,000,000 ohms. This high-resistance resistor prevents too much current from running through the circuit. Notice that the current is not flowing to the base (labeled B in the figure) of the transistor, so current won't flow through the collector (C) and emitter (E), and, therefore, the LEDs will not turn on.

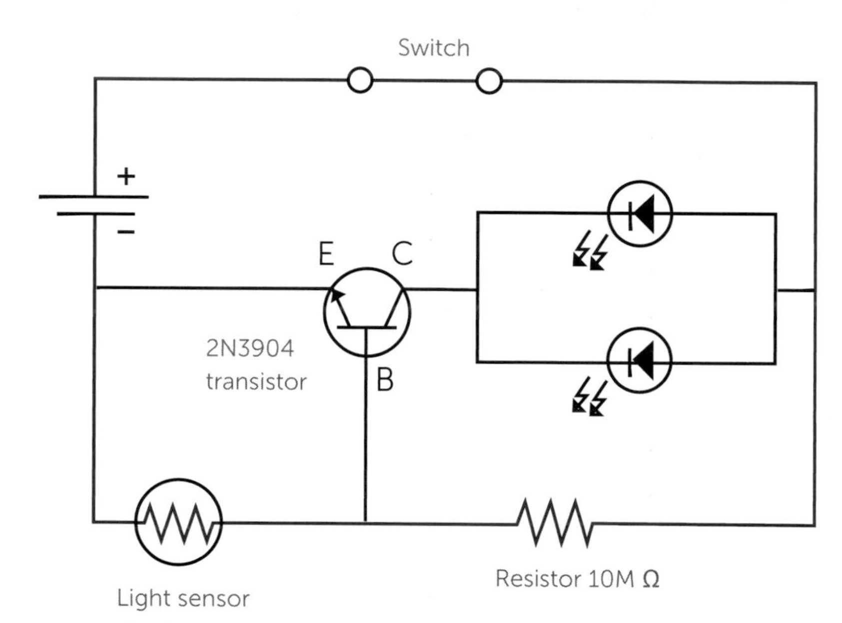

FIGURE 7.6: Dark-detecting circuit in the light

However, when the light sensor is in the dark, or if it is covered, electricity does not flow through it as easily. This makes it so the path of least resistance for the electric current leads to the base of the transistor, as illustrated by the red circuit in Figure 7.7. This triggers the transistor to allow current to flow through the collector and emitter, thus turning on the LEDs.

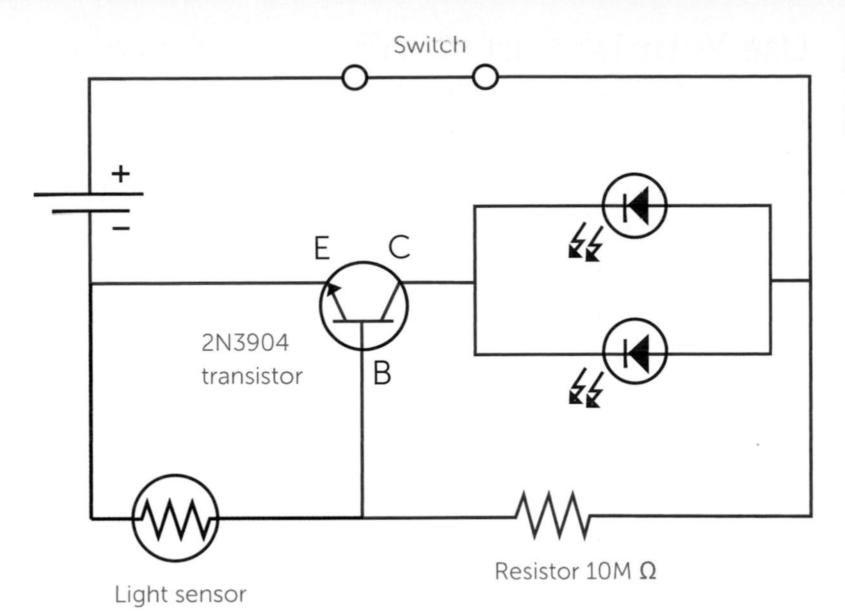

FIGURE 7.7: Dark-detecting circuit in the dark

Use Your Imagination

Now that you're familiar with the idea of how nightlights work, it's time to start this chapter's project. Get started by making a paper sketch of what you want your dark-detecting bracelet to look like. Make sure that your light sensor and LEDs are facing toward the outside of the bracelet. You should try to draw everything life-sized and make sure that your bracelet is big enough to hold all of the electronic components and your circuit (see Figure 7.8).

FIGURE 7.8: A sketch of the front of the light-sensing bracelet

Plan It Out

When you take a look at Figure 7.9, you will notice that this circuit is a bit more complicated than our previous circuits. This is why it's especially important to plan it out by drawing a life-sized circuit on the back of your paper sketch. When you do, make sure to label the positive and negative legs of your LEDs. Also label the emitter, base, and collector of your transistor. Refer to Figure 7.9 if you need help finding the emitter, base, and collector. Place all the components on the your circuit diagram before you start assembling your bracelet to make sure you haven't made any mistakes and to make sure you have enough room for all the pieces of the circuit.

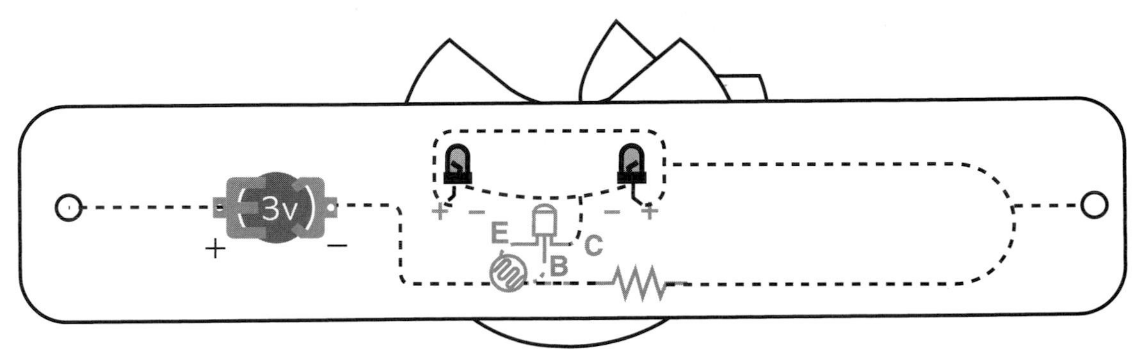

FIGURE 7.9: The circuit diagram drawn on the back of the light-sensing bracelet sketch

Preparation

After you are done sketching and carefully planning the circuit for your bracelet, you are ready to gather your materials, cut out your fabric, and sew your circuit. Remember to take your time and make sure that each stitch is neat. There are more components and electrical connections in this circuit than in the previous projects, so you have to be especially careful to make sure that you don't cross the conductive threads and that you make the knots that hold your components tight so that they don't fray and don't touch each other.

LIST OF SUPPLIES:

A Conductive thread
B Battery holder
C 3V coin cell battery
D 2 LEDs
E Light sensor (photocell)
F 2N3904 transistor
G 10M Ω resistor
H Felt sheets
I Clear nail polish
J Sewing needle
K Fabric glue
L Chalk pen
M Scissors
N Awl
O Needle-nose pliers
P Snaps

Make It

Follow these steps to make your light sensor project:

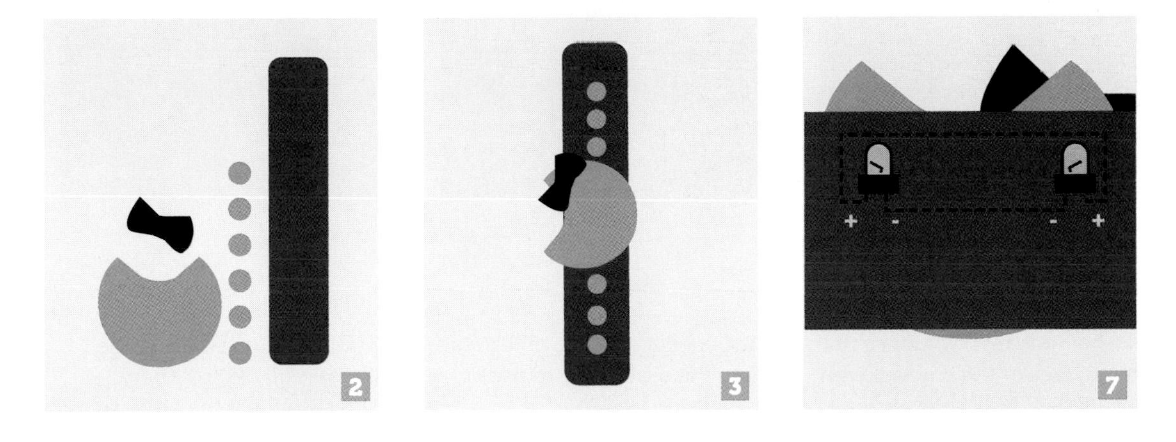

1. Sketch all of the pieces of your bracelet on a felt sheet with a chalk pen.

2. Cut out each of the felt pieces. Make sure the piece of fabric that will function as the bracelet band is long enough to go all the way around your wrist with a bit of overlap so it can snap shut and that the bracelet is wide enough to hold the entire circuit.

3. Arrange the felt pieces on the long felt bracelet shape and glue them together according to your design.

4. Using an awl, or other sharp tool, carefully poke holes for the LEDs.

5. Mark the positive and negative holes on the felt with a chalk pen.

6. Push the legs of the LEDs through the holes you created in the bracelet so that the lights are showing on the front and the legs are on the back. *Note that in these images, the lights of the LEDs are shown on the back for clarity.*

7. With conductive thread, sew the positive legs of the LEDs together and then sew the negative legs of the LEDs together.

8. Carefully poke two holes for the light sensor with your awl.

9. Insert the sensor's legs as you did with the LEDs. Again, make sure that the sensor is facing the outside of the bracelet and that the legs are on the inside. Unlike the LEDs, the sensor does not have polarity, so it does not matter which leg goes through which hole. *Again, please note that the light sensor in the illustration is shown on the inside of the bracelet for clarity. As a result, you can't see the legs in this illustration.*

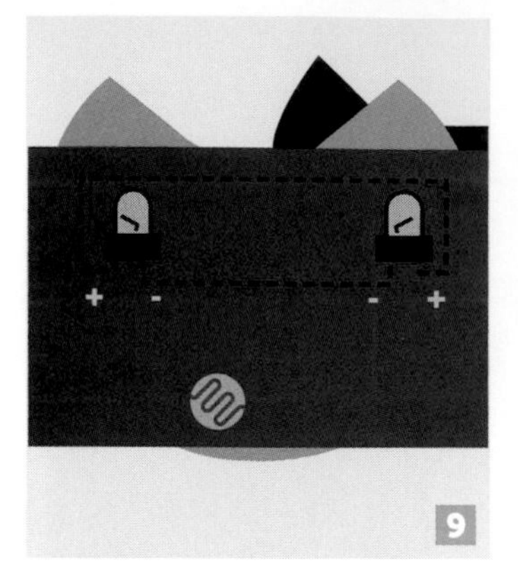

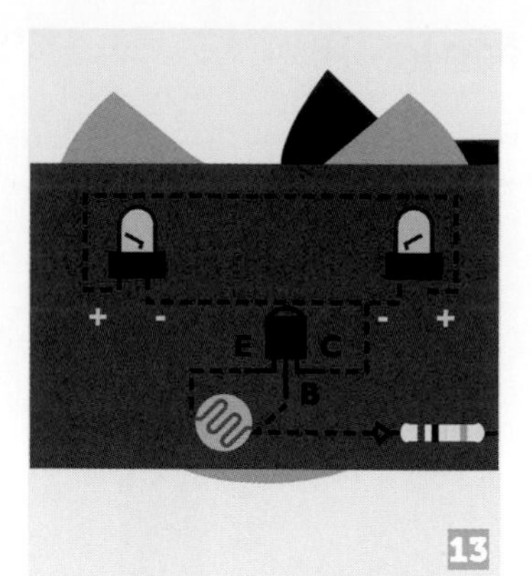

10. Draw the rest of your circuit on the back of the bracelet with a chalk pen.

The transistor can go on the inside of the bracelet, so you don't have to poke holes for the legs, but make sure it is facing the correct way, as illustrated here.

11. Connect the transistor's emitter and base to the legs of the light sensor.

12. Connect the collector of the transistor to the negative legs of the LEDs.

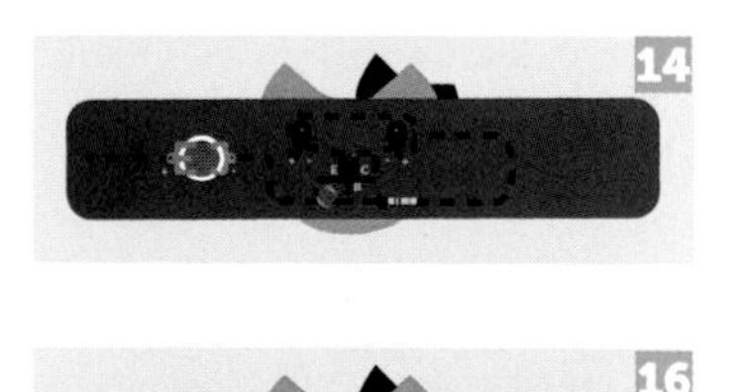

13. Connect one side of the resistor to the base of the transistor.

14. Glue a battery holder onto the bracelet. Make sure it is positioned with the negative side pointed toward the circuit. Connect the negative side of the battery holder to the transistor's emitter and one of the legs of the light sensor. Connect the loose end of the resistor to the positive legs of the LEDs.

15. Attach a metal snap fastener to each end of the bracelet. Make sure they are on opposite sides of the bracelet. *In the following illustrations, they are shown on the same side for clarity.*

16. Connect one snap to the positive side of the battery holder. Connect the other snap to the connection between the resistor and the positive legs of the LEDs. These snaps will serve as a switch that turns the circuit on and off.

17. You're almost done! Test the circuit by putting in a battery, closing the snaps, and covering the sensor. The eyes should light up.

18. After you make sure the circuit works, sew a small sheet of felt over the exposed circuit, to prevent your skin from touching the circuit, but make sure you can still change the battery.

16

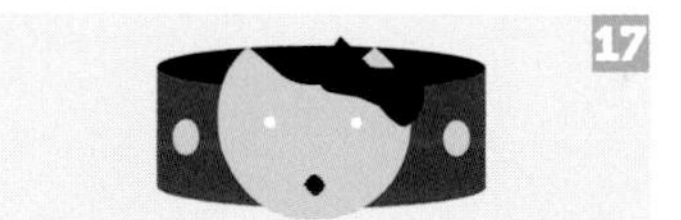

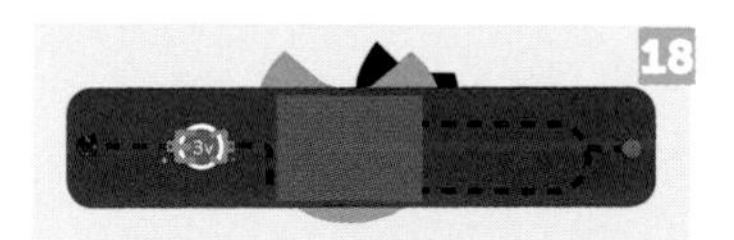

Figures 7.10 and 7.11 show the completed bracelet.

FIGURE 7.10: The finished bracelet

FIGURE 7.11: The exposed circuit on the back of the bracelet

Now that you've finished the bracelet, can you make up a story about why the cat's eyes light up when it gets dark?

Have you ever noticed that cat's eyes reflect light at night? This is because the back of cat's eyeballs have a layer of tissue called the *tapetum lucidum* that is reflective.

Take a look at some other project ideas that use this same circuit. Figure 7.12 shows my daughter modeling a similar bracelet.

Did you know that owls' eyes also reflect light at night? Like the cat bracelet, I made the owl in Figures 7.13 and 7.14 so that its eyes light up when its beak is covered.

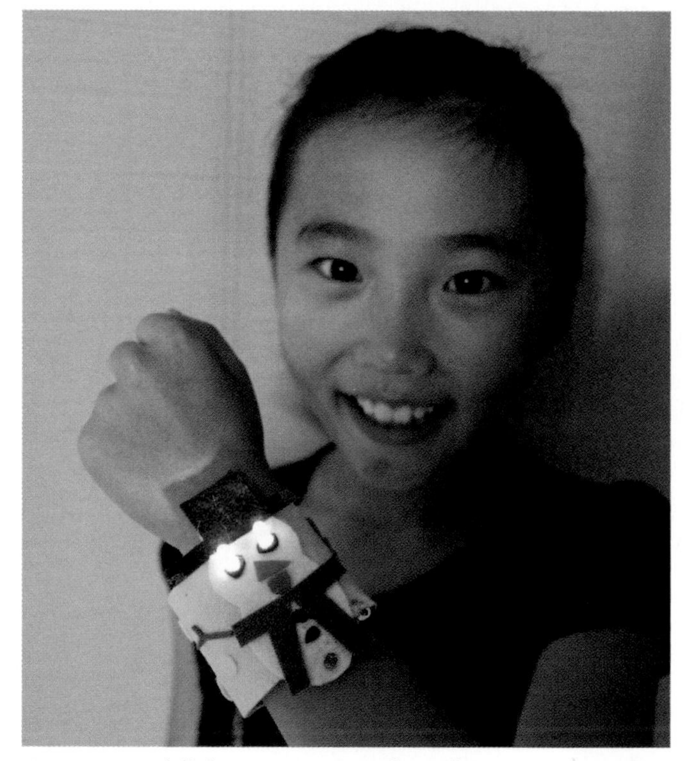

FIGURE 7.12: A light-up snowman bracelet

FIGURE 7.13: The night-guarding owl

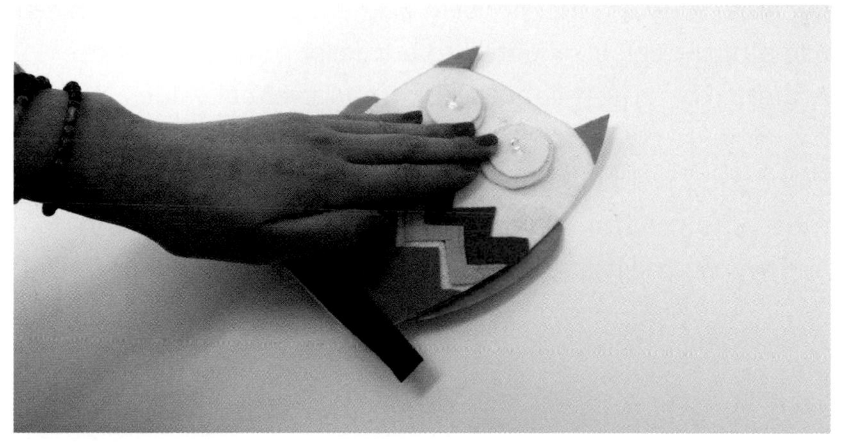

FIGURE 7.14: The night-guarding owl with sensor covered and eyes lit up

Figure 7.15 shows another light-up project idea. You cover the sensor at the bottom to make the lights at the top of the crown turn on.

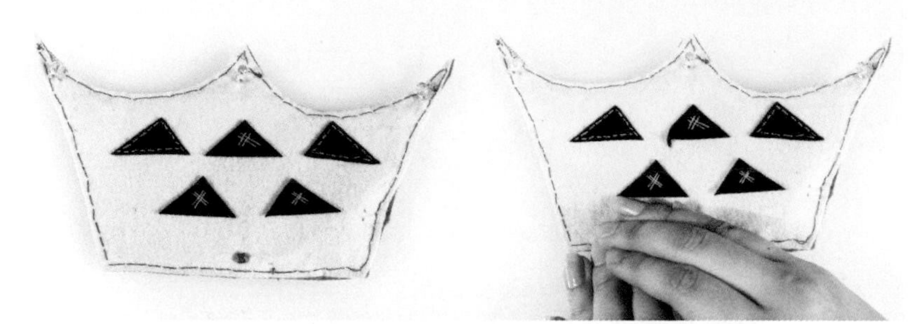

FIGURE 7.15: A crown with points that glow in the dark

Now that you've seen several examples of projects, do you think you can use this circuit to make a T-shirt or a bag that lights up in the dark?

Remember, light sensors, or photocells, are made up of materials that change their electrical resistance according to how much light is shining on them. They are common because they are inexpensive and easy to use. You can use a photocell connected to a microcontroller like the Arduino to convert light levels into numerical values. You can then go on to use those sensor values in a program like Processing to make interesting interactive art. To learn more about Arduino and Processing, visit www.arduino.cc and www.processing.org.

Using Sensors to Control a Computer

I was very excited when I first learned that I could control a computer with signals from a sensor. The first project that I made was called "Be Happy." It used two photocells to control sound and the images on a computer screen. Figure 7.16 shows the circuit prototype of Be Happy on a breadboard, a device that makes it fast and easy to make and test circuits. Figure 7.17 shows the completed Be Happy project that has an interface with two faces and a computer screen.

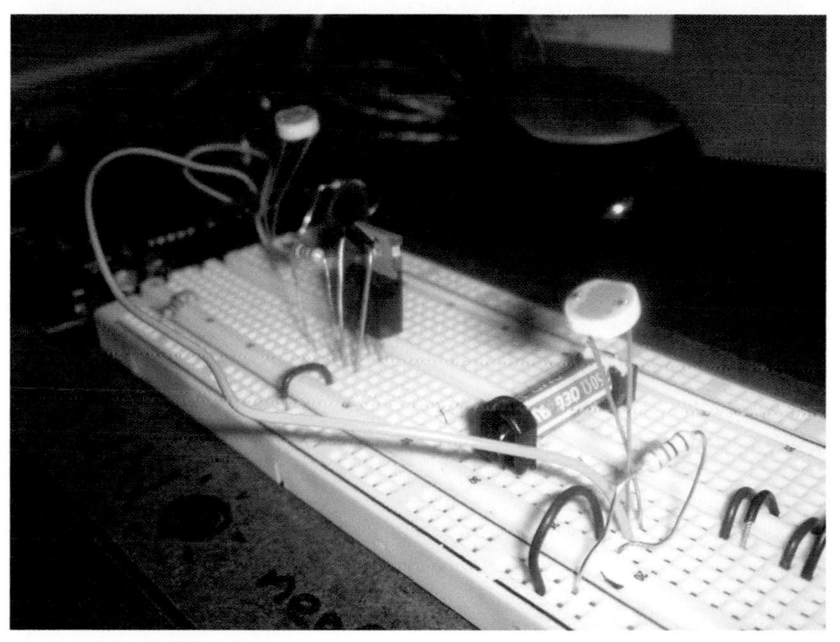

FIGURE 7.16: The Be Happy circuit on a breadboard

FIGURE 7.17: Be Happy interface

Using Be Happy

Take a look at the following image. As you can see, when both faces are uncovered, a sharp red image appears on the monitor.

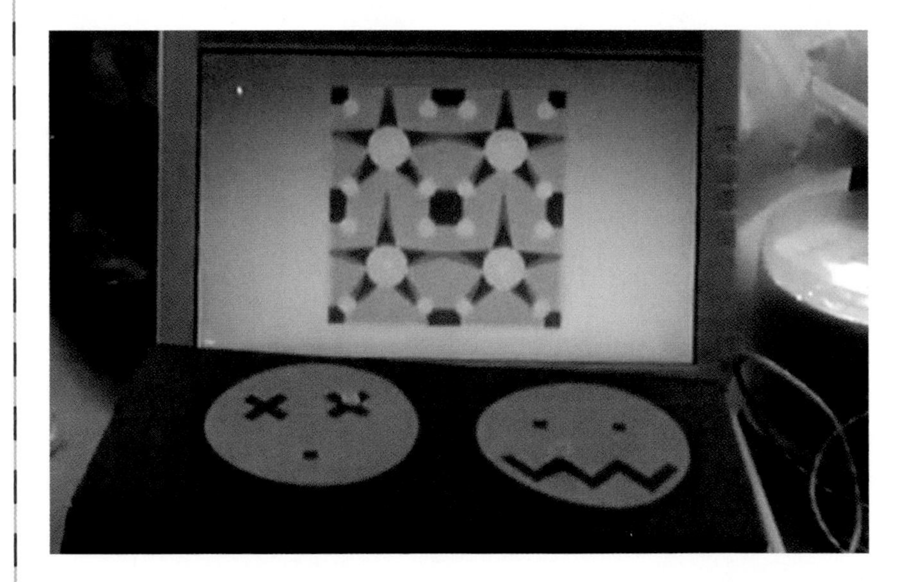

However, if you cover the left face, the image becomes more round and bright.

If you cover the right face, on the other hand, the image doesn't change but a fun and cheerful sound comes from the computer.

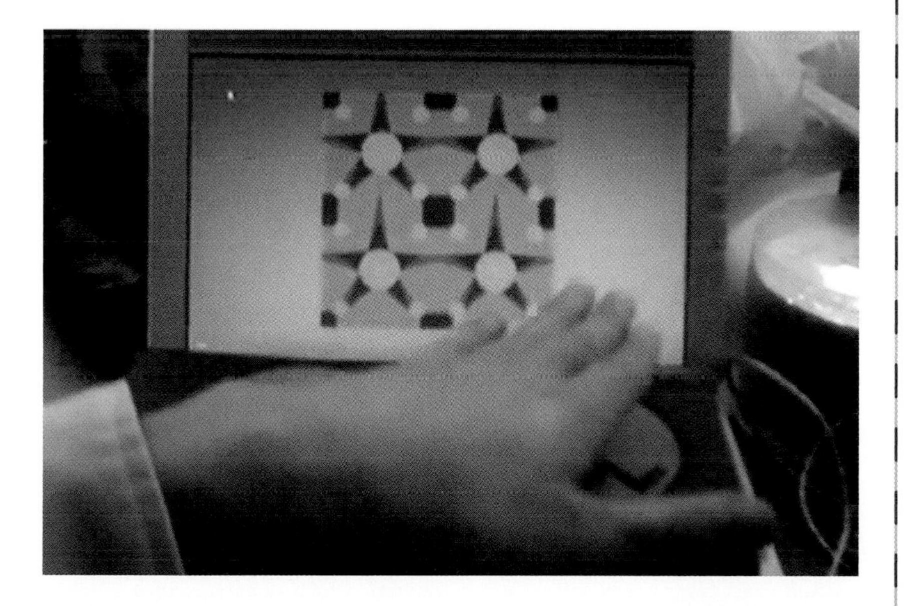

If you cover both faces, both the previous changes occur; that is, the image on the screen becomes more round and bright *and* a fun, cheerful sound plays from the computer.

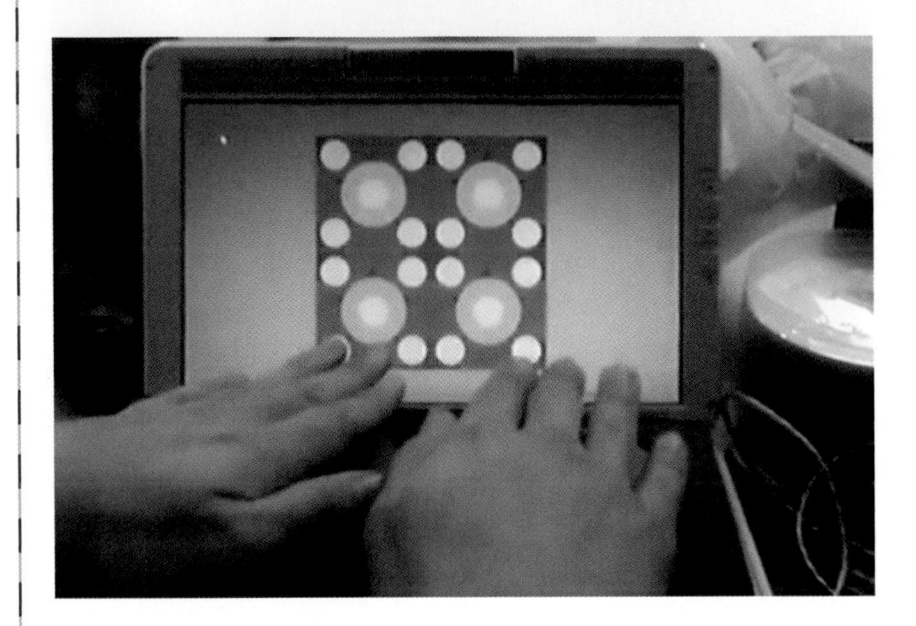

You can watch a video of the Be Happy project here: **http://goo.gl/L80utZ**.

What's Next?

Nice work! Now that you've learned how to use a transistor and a light sensor to make a nightlight circuit, in the next chapter, you are going to learn a bit about integrated circuits and how we can use them to create more complex behaviors in our circuits. You will make a project that blinks LEDs and beeps a buzzer all by itself.

Chapter

8

A Beeping, Blinking UFO

USING THE 555 TIMER

Now that you know how to turn on and off light-emitting diodes (LEDs) using switches and light sensors, you are ready to learn how to make an LED blink on and off all by itself. For this project, I am going to introduce a new type of electronic component called an integrated circuit (IC). An integrated circuit, commonly called a microchip, is just a tiny circuit inside a plastic case. You've probably seen ICs inside electronics or computers. For this project, you will use a very popular IC called a 555 timer to make a blinking and beeping UFO.

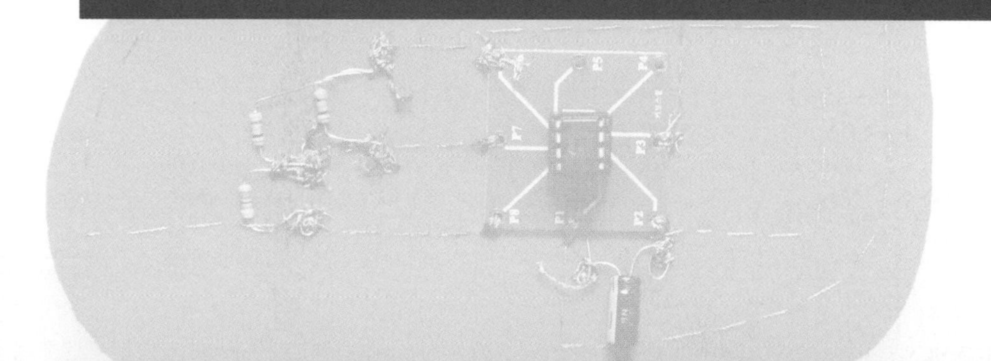

To make this project, you are going to have to learn a bit more about integrated circuits and how to use them. You will also learn how to use another electronic component called a *capacitor*.

Integrated Circuits (ICs)

An *integrated circuit* is really just a tiny circuit that includes several electronic components like resistors and transistors inside of a plastic case. You can connect to this circuit through the metal legs, called *pins*, that come out of the case. Many types of integrated circuits can perform many different functions. Some ICs act as counters, amplifiers, memory storage devices, microprocessors, and more.

555 Timer

The IC that we are using for this project is called a 555 timer (see Figure 8.1). The 555 timer is a popular chip with many

FIGURE 8.1: A 555 timer
Photo credit: https://upload.wikimedia.org/wikipedia/commons/thumb/6/64/NE555_DIP_%26_SOIC.jpg/1280px-NE555_DIP_%26_SOIC.jpg

uses. This IC is a type of *oscillator*, or a circuit that allows you to vary the voltage coming from one of the pins on the IC in a wavelike pattern. If you can get the voltage to go up and down in a regular pattern, then you can use that changing voltage to get an LED to blink on and off. The 555 timer can generate electrical pulses, or oscillations, with intervals from over one hour to only 0.00005 seconds! The 555 timer has eight pins to which you can connect your circuit. To use any IC, you need to know how to connect it to your circuit. For this, you can use a *pinout diagram* like the one shown in Figure 8.2.

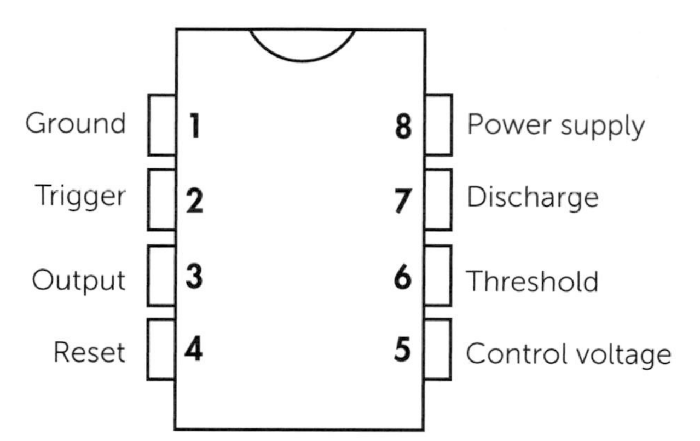

FIGURE 8.2: The 555 pinout diagram shows the function of each pin on the 555 timer.

An important note about sewing circuits with integrated components: The legs, or pins, on ICs are very small and close together. It would be really difficult to sew directly to them. As a solution, you can use a specially designed circuit board called a *breakout board*. When you plug a 555 timer into a breakout board, then the pins of the IC are connected to holes that spread out, and these are easy to connect to with your conductive thread (see Figure 8.3). For a list of suppliers for this part, please visit the website for this book at **www.techdiy.org**.

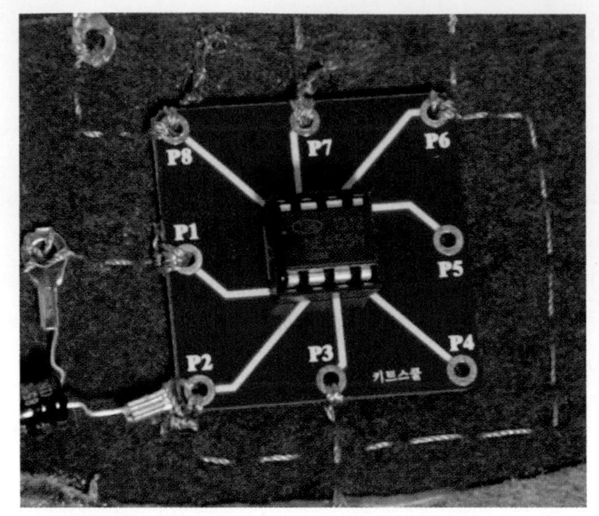

FIGURE 8.3: 555 timer on a breakout board

Capacitors

A capacitor is used to store electrical charge in a circuit. Similar to the way your body builds up static electricity in the winter and then releases it when you touch a doorknob, a capacitor allows the electrical energy between two separated plates to build up until it is so great that the electricity jumps across the plates, discharging the capacitor in the process. Capacitors are helpful for absorbing and stabilizing voltage as well as for charging and discharging electricity as a battery does (see Figure 8.4). The amount of charge that a capacitor can hold is measured in farads. Common capacitors store microfarads for softening voltage changes or for controlling short intervals of electrical signals. A super capacitor has 1,000 times more storage capacity than a common capacitor.

The Circuit

The circuit diagram in Figure 8.5 shows how to blink two LEDs using a 555 timer, a 10 microfarad (uF) capacitor, a 200 ohm resistor, and a 10,000 ohm (10k ohm) resistor. The 10k ohm resistor, shown here between pins 6 and 7, is used to control the rate of blinking. If you increase the size of that resistor, the blinking will slow down. As shown in Figure 8.5, you can also add a buzzer in a parallel circuit with two LEDs to make your UFO beep as well as blink.

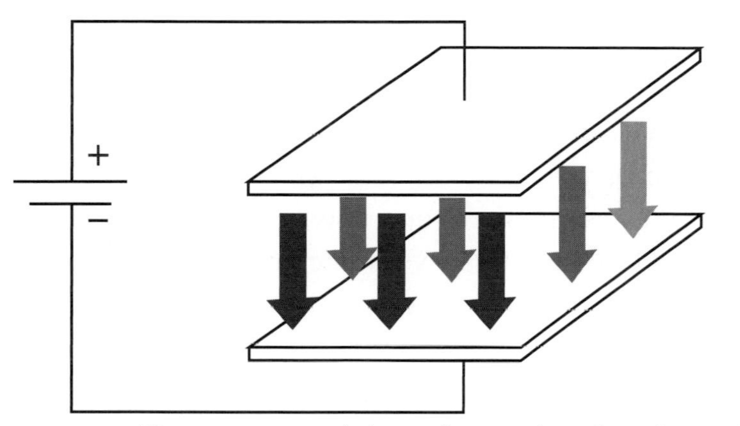

FIGURE 8.4: The two separated plates of a capacitor allow electrical charge to build up until it is so great that the electrons jump across the gap between the plates.

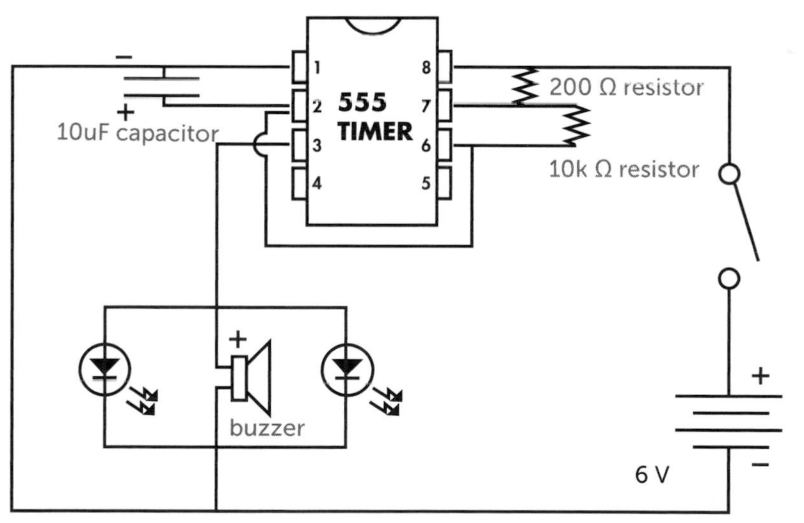

FIGURE 8.5: The 555 timer circuit for blinking LEDs and beeping a buzzer

Use Your Imagination

The circuit in Figure 8.5 makes two LEDs blink and a buzzer make a beeping sound. What suitable project ideas can you come up with that blink and beep? Use your imagination. In our example project, we are going to make a UFO. Make a life-size sketch of your project like the one in Figure 8.6.

Plan It Out

Once you have decided on your design and sketched it out, check out the circuit diagram in Figure 8.7 to decide where to position your components and how to connect them with conductive thread. Draw your circuit on the back of your project sketch. Give yourself plenty of room because this is a complicated circuit with many components and connections. It is a good idea to place the 555 timer near

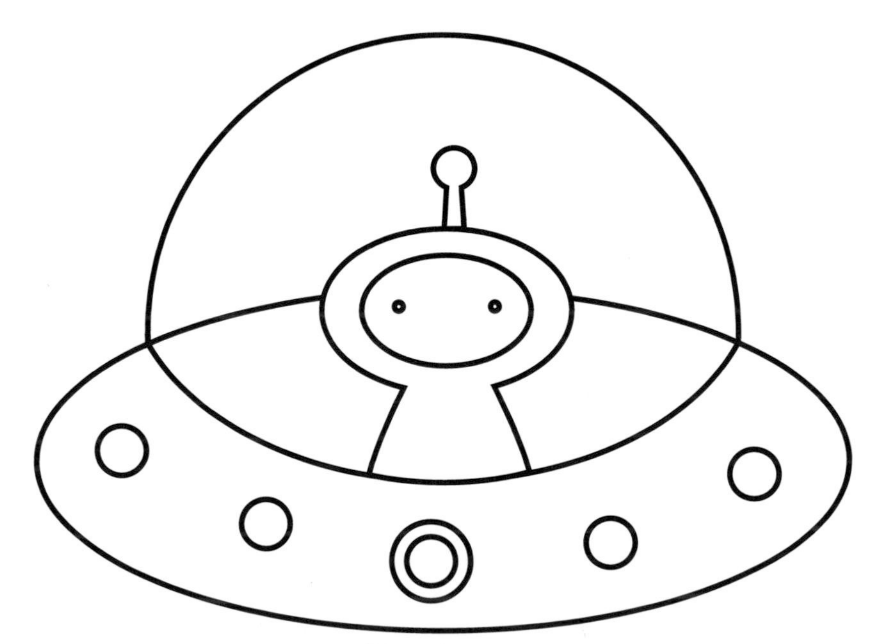

FIGURE 8.6: The front of our UFO project

the middle of your circuit. You can then put the rest of your components on your circuit diagram to make sure you haven't made any mistakes and that you have enough room for all the pieces of the circuit.

Preparation

After your have sketched and planned out your project, you are ready to gather your materials and begin making it. Be aware that as your circuits get more complicated, with more components and more connections, you really have to be careful that you don't cross the conductive threads; you also need to make sure that the knots that hold your components are tight, that they don't fray, and that they don't touch each other.

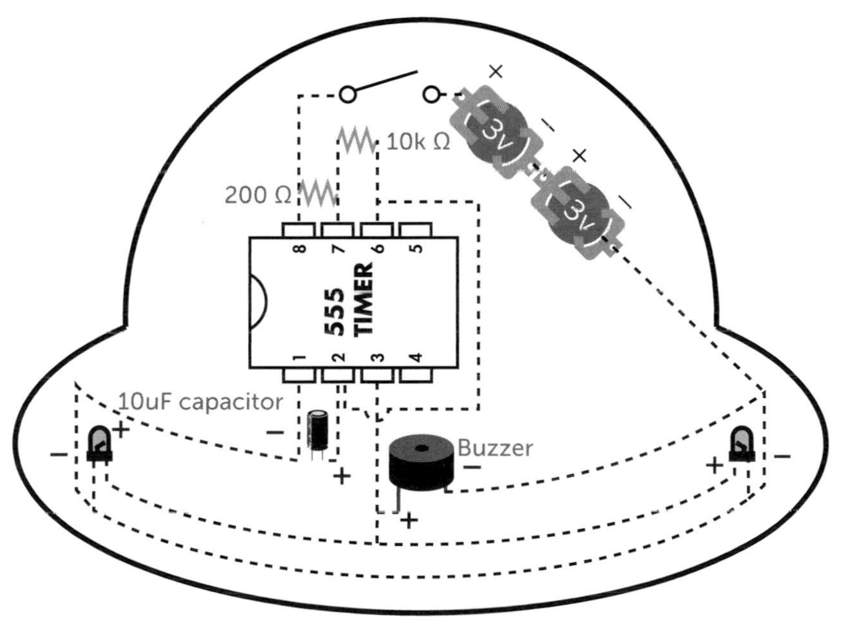

FIGURE 8.7: The circuit diagram for our UFO project

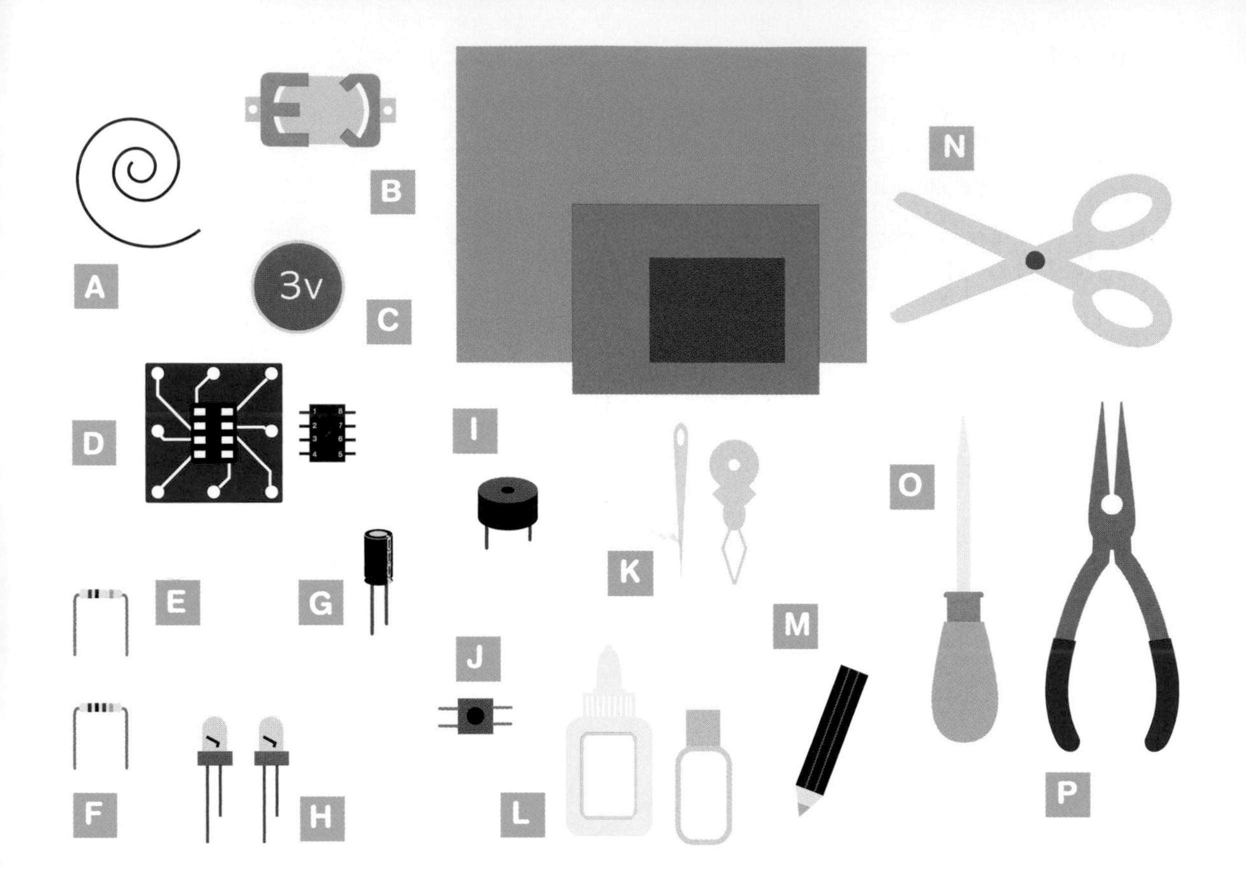

LIST OF SUPPLIES:

- A Conductive thread
- B Battery holder
- C Coin cell battery
- D 555 timer and breakout board
- E 10k ohm resistor
- F 200 ohm resistor
- G 10uF capacitor
- H 2 LEDs
- I Buzzer
- J Push button
- K Needle and needle threader
- L Glue and clear nail polish
- M Chalk pen
- N Scissors
- O Awl or other sharp tool for poking holes in fabric
- P Needle-nose pliers

Make It

Follow these steps to make your timer circuit:

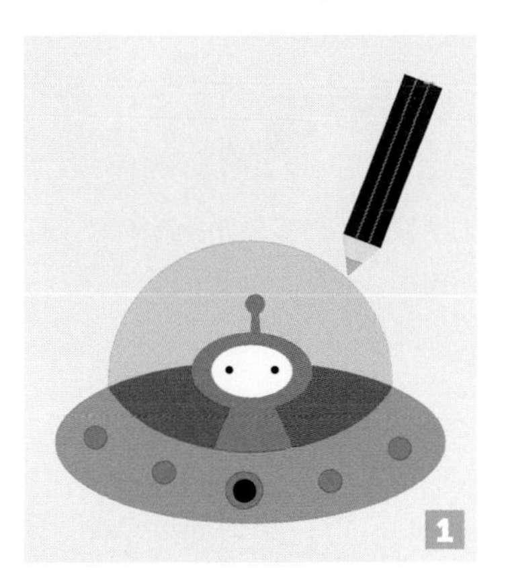

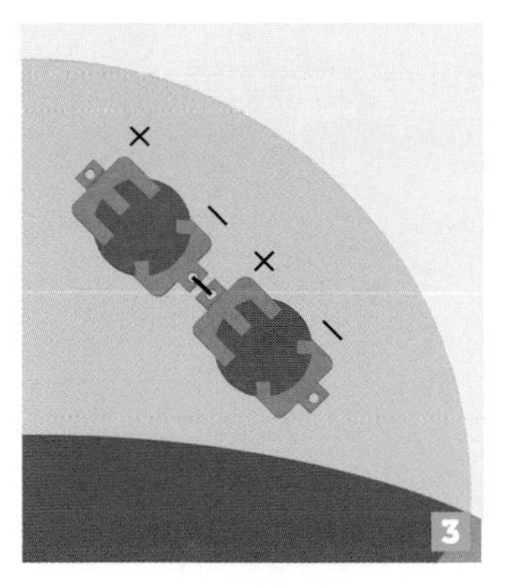

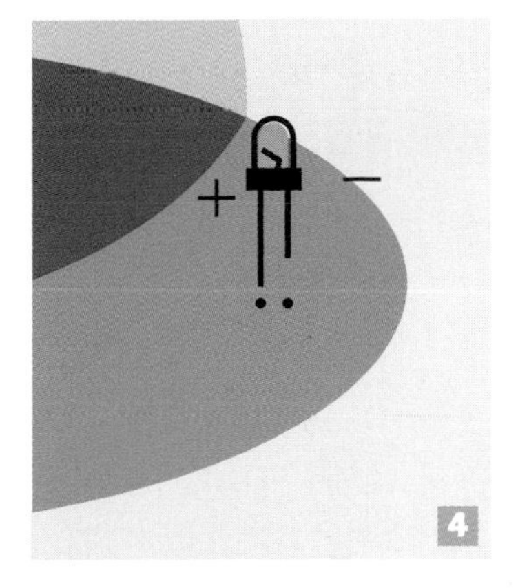

1. Use a chalk pen to sketch the pieces of felt according to your design.

2. On the back of your felt, glue the two battery holders in a series. Make sure that the positive and negative sides are pointing in the correct directions.

3. Using conductive thread, connect the two battery holders together in the middle.

4. Figure out where you want to put your LEDs and buzzer. Mark the positive and negative holes for your LEDs.

5. Poke the legs of the LEDs and the buzzer through the felt.

6. Using conductive thread, connect the positive legs of the LEDs to each other. Do the same with the negative legs.

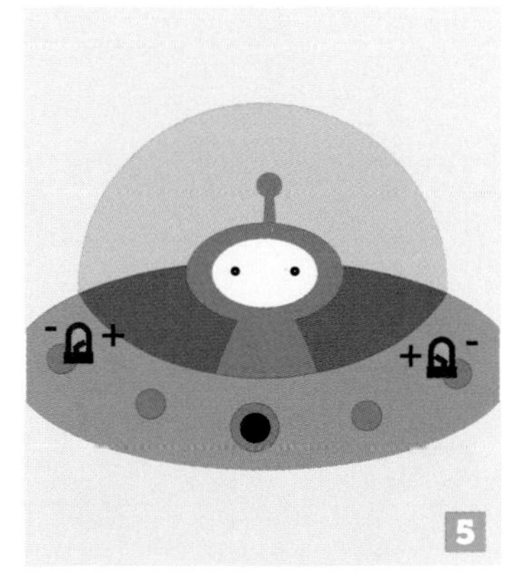

7. Sketch the circuit diagram on the backside of the project with a chalk pen. Mark the positions of the positive and negative legs for the capacitor between pins 1 and 2 on the 555 timer. Also mark the positive and negative legs for the buzzer.

8. Carefully start to sew your circuit with conductive thread.

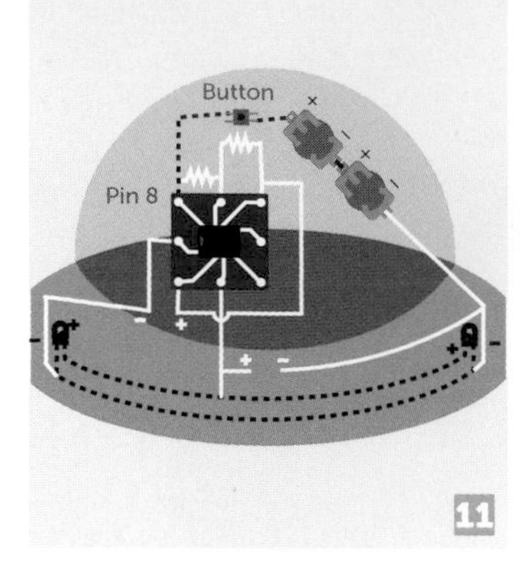

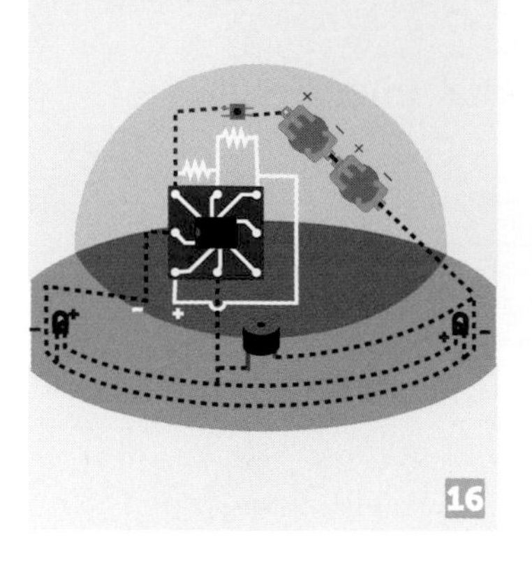

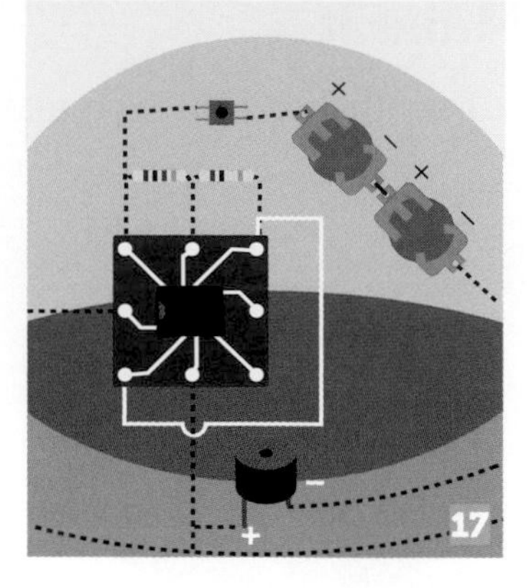

As you begin each thread, double-check the circuit diagram in Figure 8.5 to make sure that you are connecting the components properly.

9. Start by connecting pin 8 on the 555 timer to one of the legs on a push button. These buttons have four small legs. You can use any two that are diagonal from each other.

10. Use needle-nose pliers to bend the legs into hooks so that you can sew to them.

11. Finish attaching the button by connecting the diagonal leg to the positive side of the battery.

12. Now add the buzzer. The buzzer has two legs, the longer of which is the positive leg. You'll have to use needle-nose pliers to carefully bend the buzzer legs into hooks so that you can sew to them.

13. Connect the positive leg of the buzzer to pin 3 on the 555 timer. Also connect pin 3 on the 555 timer to the positive legs on your LEDs.

14. Connect the negative (shorter) leg of the buzzer to the negative legs of the LEDs.

15. Connect the LED's negative legs to the battery holder's negative pole.

16. Then connect the negative legs of the LEDs to pin 1 on the 555 timer.

17. Add the two resistors to your circuit. Connect pins 6 and 7 on the 555 timer through a 10k Ω resistor and connect pins 7 and 8 through a 200 Ω resistor.

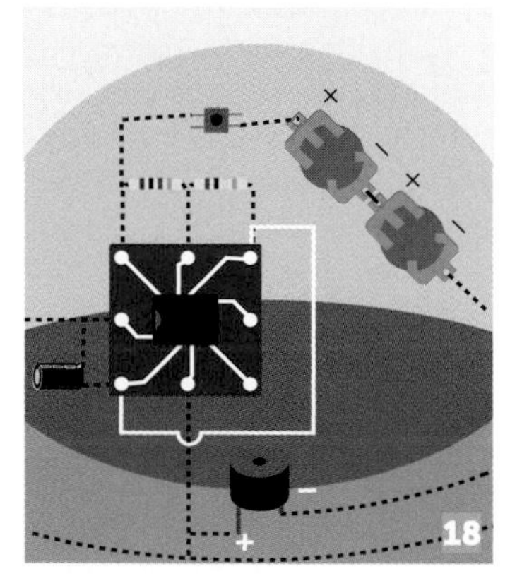

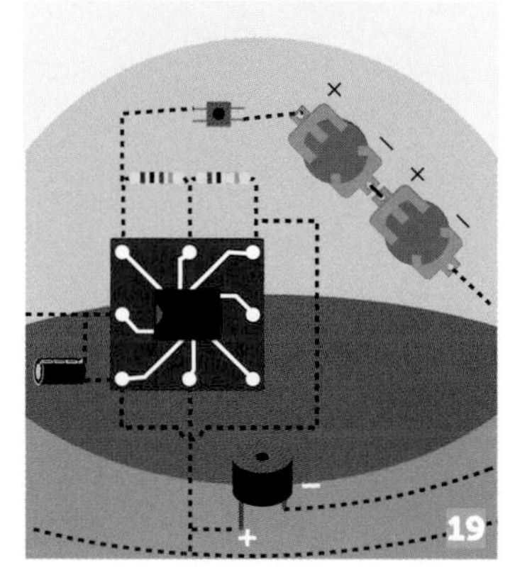

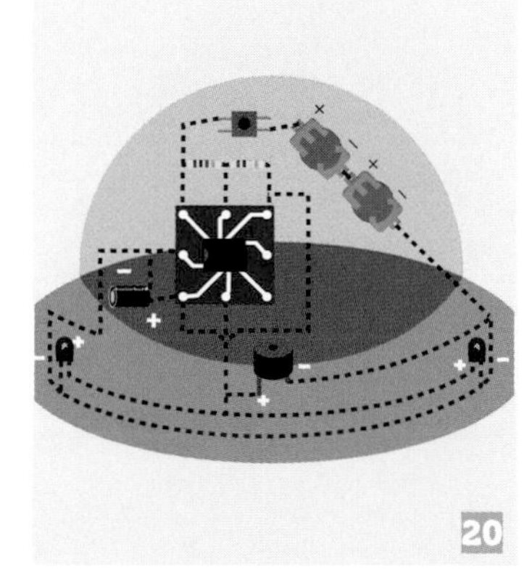

18. Add the capacitor to your circuit. Connect the positive (longer) leg of the capacitor to pin 2 on the 555 timer. Then connect the negative (shorter) leg of the capacitor to pin 1 on the 555 timer.

19. Finally, connect pins 2 and 6 on the 555 timer. To do so, you are going to have to jump over the thread that is connected to pin 3. The bump in the illustration indicates that the two threads do not touch. Be very careful to avoid touching these two threads together. This would cause a short circuit. Make sure that the two threads cross on opposite sides of the felt.

20. Once you're done connecting the pins, double-check that all your parts are connected and in the right places by carefully comparing the circuit diagram in Figure 8.5 and your circuit, which should look similar to this illustration.

Start with pin 1 on the 555 timer. Make sure it is connected to all of the correct components and that the components are facing in the correct directions. Then move on to pin 2 and do the same thing. Repeat this process for all eight pins on the 555 timer. Note that pins 4 and 5 are not connected to anything. That is okay. We are not using those pins in this circuit.

21. Finally, insert the batteries in the battery holders. If everything is connected properly, when you hold down the push button, you should see the LEDs flashing and hear a beeping sound!

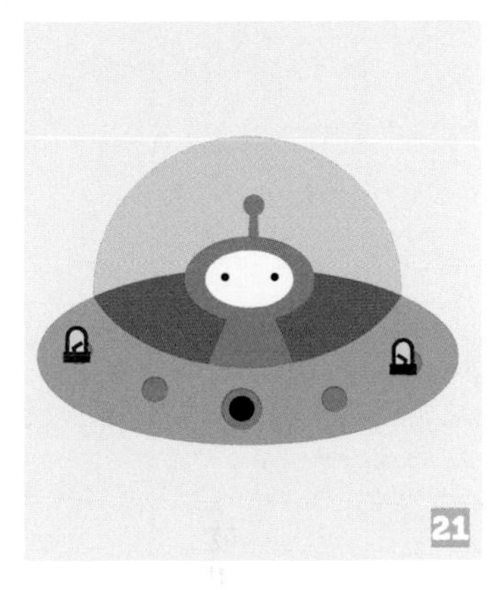

Figure 8.8 and 8.9 show both sides of the completed project.

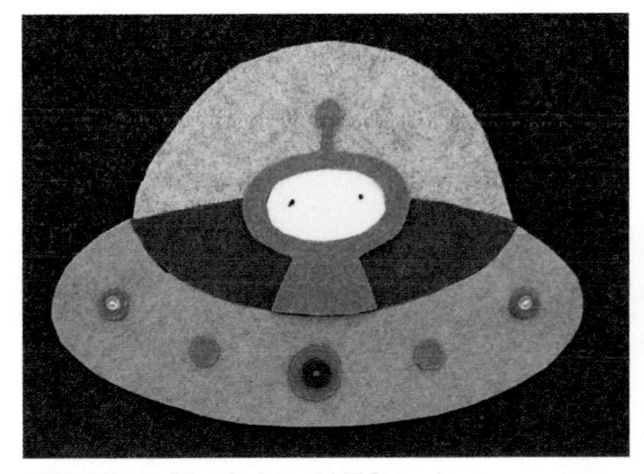

FIGURE 8.8: The finished UFO project

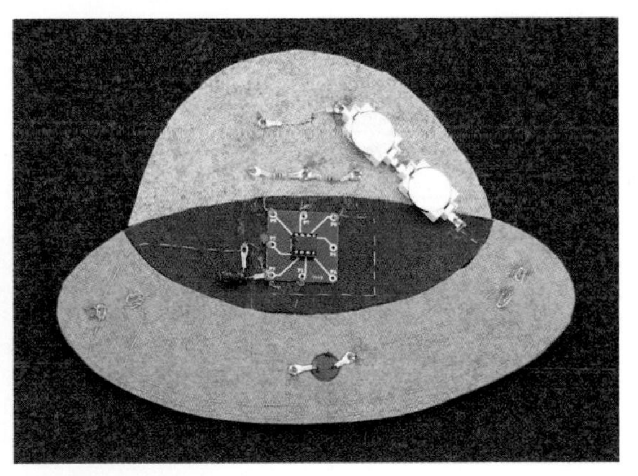

FIGURE 8.9: The finished UFO circuit

The 555 timer is a popular chip for electronics enthusiasts because it is inexpensive, easy, and safe to use. You can make a lot of different projects with this chip.

Figures 8.10 and 8.11 show another idea for a project that uses a 555 timer.

FIGURE 8.10: This Christmas stocking has twinkling lights on a tree and snowman. You could also add this circuit to ornaments for your tree.

FIGURE 8.11: The 555 timer circuit on the back of the stocking

Figures 8.12 and 8.13 show a police car that I made that makes a beeping sound and has some flickering lights. I found that if the buzzer is too loud, I could cover it with some tape to dampen the sound.

FIGURE 8.12: A friendly looking police car

FIGURE 8.13: The circuit on the back of the police car

Prototyping Circuit Boards

Now that you've seen some project ideas, are you curious about how you can make your own circuit boards like the breakout board for the 555 timer? You can prototype your own circuit boards using something called a *breadboard*. A breadboard like the one in Figure 8.14 allows you to quickly and easily connect electronic components together to test your circuit.

Once you are sure that your circuit is working with the breadboard, you can design your circuit on a computer using a circuit design program like Eagle, Fritzing, or 123D Circuits. Both Fritzing and 123D Circuits allow you to order circuit boards directly from their programs, or you can send your circuit designs to a company like OSH Park that prints and mails circuit boards. Or if you're even more

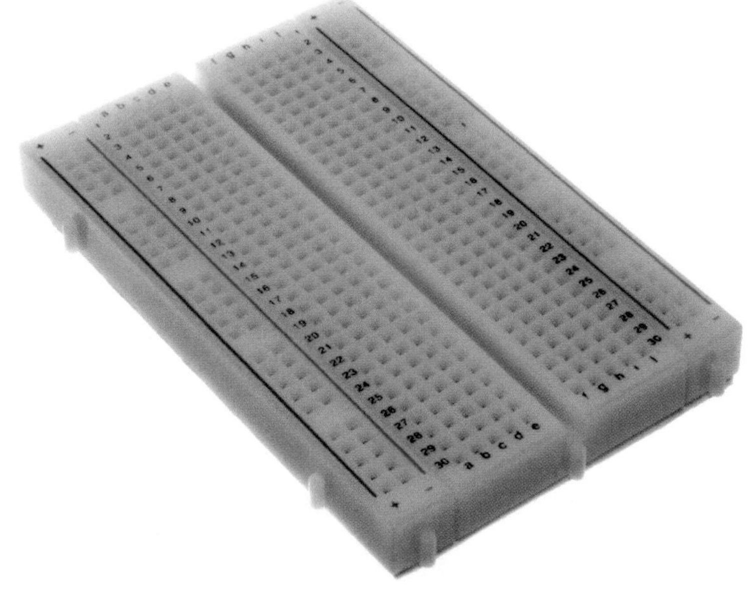

FIGURE 8.14: A breadboard

adventurous, you can make your own circuits using a chemical etching process. To learn more about this process, check out this tutorial: http://makezine.com/projects/pcb-etching-using-toner-transfer-method.

The Tales of Grimm Dollhouse

Figures 8.15–8.17 show pictures of a school project for which I made my own circuit boards. This dollhouse has motorized puppets in each room. By turning the pages in a book, the reader controls the puppets and causes them to act out scenes from different fairy tales.

FIGURE 8.15: The tales of Grimm dollhouse

Figure 8.16 shows the custom-made circuit board I designed for the dollhouse. All of these wires are connected to the sensors on the pages of the book and different motors control the different puppets.

Figure 8.17 shows the custom circuit board I designed for this project.

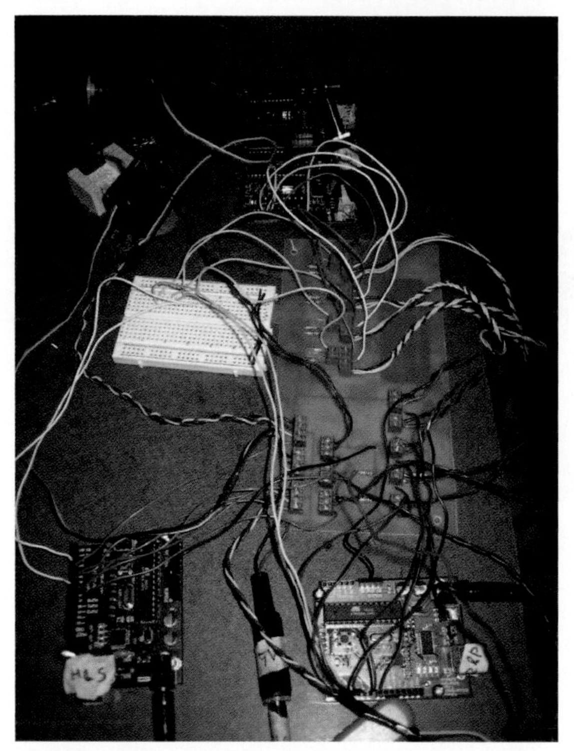

FIGURE 8.16: The complicated circuit for the dollhouse project

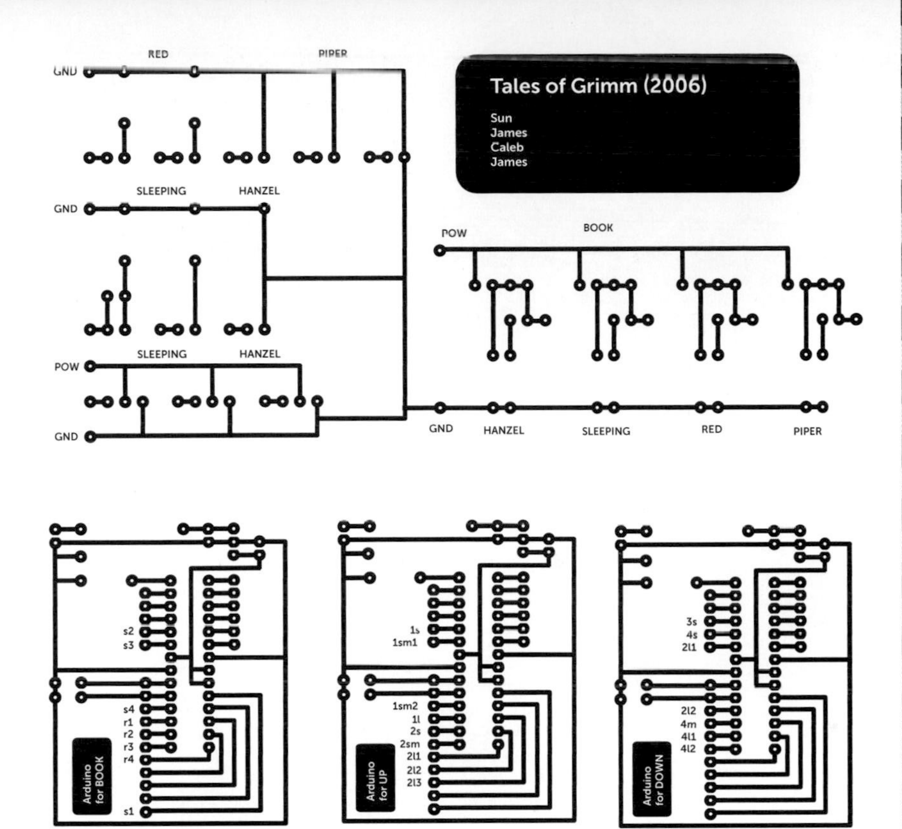

FIGURE 8.17: The design for the custom circuit board

What's Next?

Wow! Now that you've learned how to use the 555 timer to make LEDs blink and a capacitor to store electricity, the next chapter is going to free you from batteries! You will use a solar panel and a super capacitor to collect, store, and use the energy from the sun!

Chapter
9

Solar Sun

WORKING WITH SOLAR POWER

All of the projects that you have made so far have used coin cell batteries for their power source. But once those batteries run out, you'll have to dispose of them somehow. Usually that means recycling them or throwing them away, which is not a sustainable way to power your projects because the batteries contain dangerous metals that can get released into the environment.

A better solution is to use a solar panel for power. Because solar panels use renewable energy from the sun, they don't run out of electrical charge like batteries, so you don't have to recycle them or throw them away. In this project, you are going to learn how to use a solar panel and a super capacitor to power light-emitting diodes (LEDs).

Let's talk about how solar panels work and how you can use super capacitors to store energy from the sun and power your projects with renewable energy.

Solar Cells

Did you know that even though the sun is 93 million miles away from the earth, it still emits enough energy *every second* to supply the power needs of our planet for almost 500,000 years? If only we could harness all of this energy!

A solar panel, or solar cell, can convert the light energy from the sun into electricity (see Figure 9.1). A solar panel contains two types of a material called silicon. The top sheet is called *n-type silicon* and it contains extra electrons. The bottom sheet is called *p-type silicon*. This material contains "holes" where electrons would like to go. When the sun shines on the top of the solar panel, the light knocks into the extra electrons in the n-type silicon and causes the electrons to flow across to the p-type silicon. This flow of

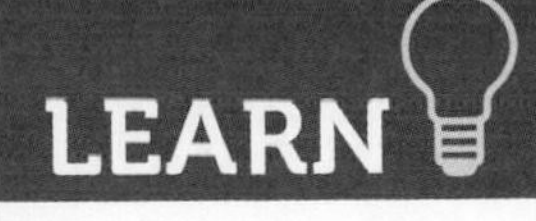

electrons creates an electrical current that you can use to power your projects!

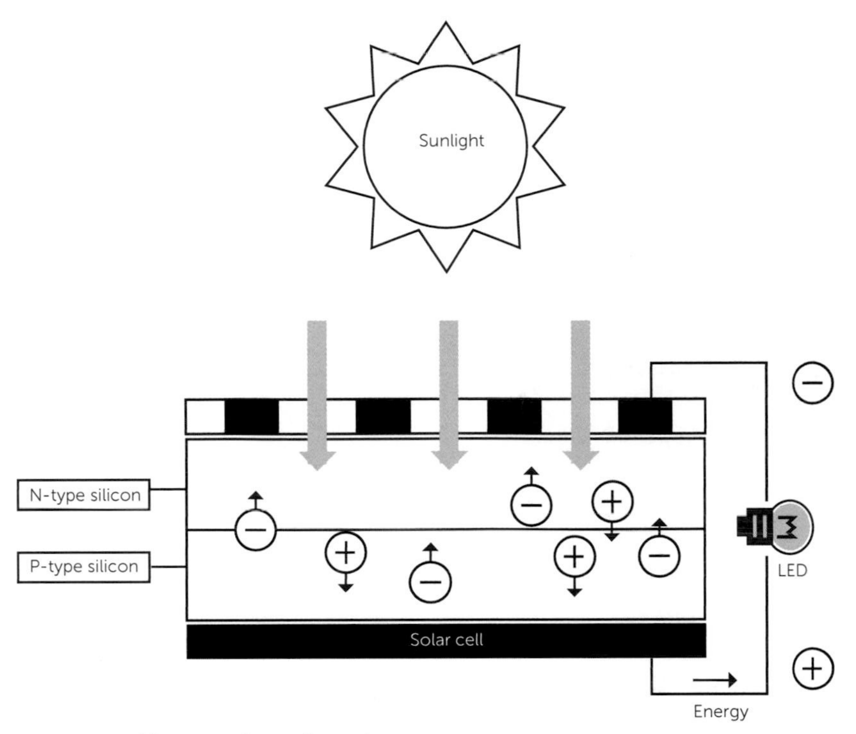

FIGURE 9.1: How a solar cell works

The Circuit

Although the solar panel provides the power for this circuit, this power needs to be stored somewhere. For this project, you will store the electricity in a *super capacitor*. This is just like the capacitor that you used in the UFO project in the previous chapter, except it can store even more electrical charge. You'll also have to include a diode in the circuit as well. Electricity can only flow one way through a diode, so this will prevent electricity from moving backward through your circuit and leaking out of the solar panel (see Figure 9.2).

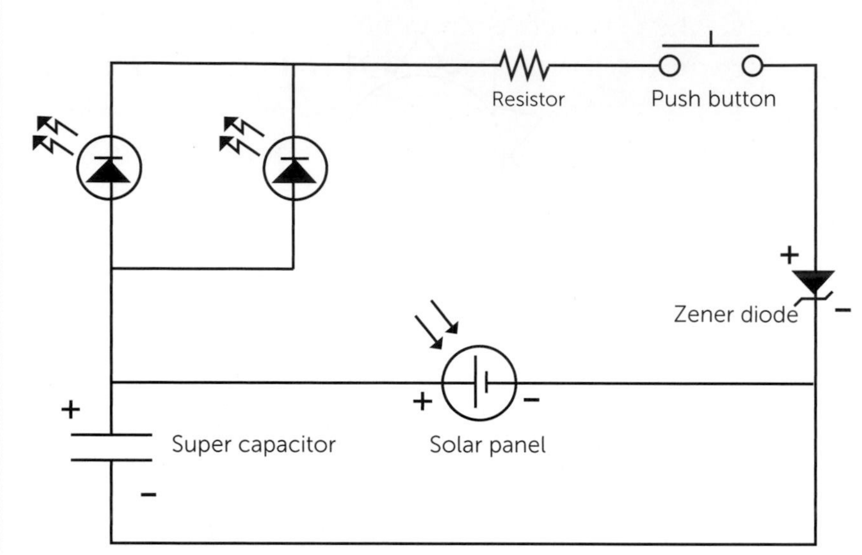

FIGURE 9.2: The solar panel circuit

In full sunlight, the solar panel in this circuit provides 6 volts (V) at 160 milliamps (mA), but the super capacitor and the diode will use up some of that voltage, so the voltage available for your LEDs is actually closer to 4.5 volts. When choosing a solar panel, it is usually a good idea to pick one with a higher voltage than what you think you need. For recommendations and suppliers, please visit the website for this book at www.techdiy.org. Also keep in mind that a solar cell that emits a higher current will be able to charge your super capacitor even faster than one that emits a lower current.

Use Your Imagination

Before you start your drawing, consider the size of your solar panel and where it will fit on your project. Don't forget that your solar panel only generates electricity when it is exposed to light, so make sure that you don't cover it up! Make a life-sized sketch of your project similar to the one in Figure 9.3.

FIGURE 9.3: A sketch of the Solar Sun project

Plan It Out

In your drawing, include the actual sizes of your solar panel and the other components. And just as you have done with your previous designs, make sure that you don't put your components too close together. You want to give yourself room to sew your connections without getting your conductive thread tangled or crossing it between components. Figure 9.4 shows an example of what such a drawing should look like.

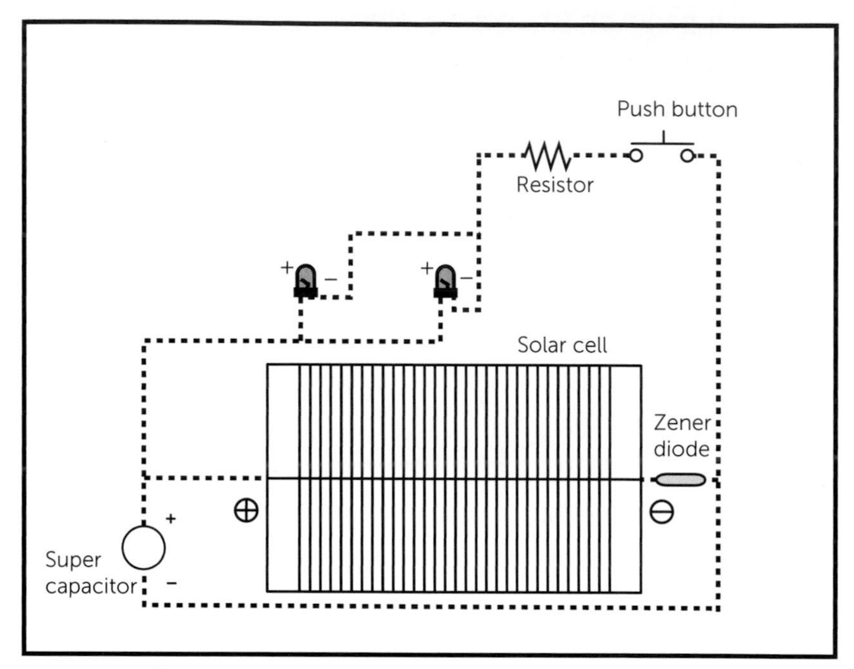

FIGURE 9.4: The back of the Solar Sun project with the circuit diagram

Preparation

Once you are done sketching and planning out your project, you can collect your materials and begin putting everything together. Just as with the previous projects, go slowly. Focus on making each stitch neat and complete before moving on to the next stitch. Also, double-check that your components are facing the right direction before you sew them down.

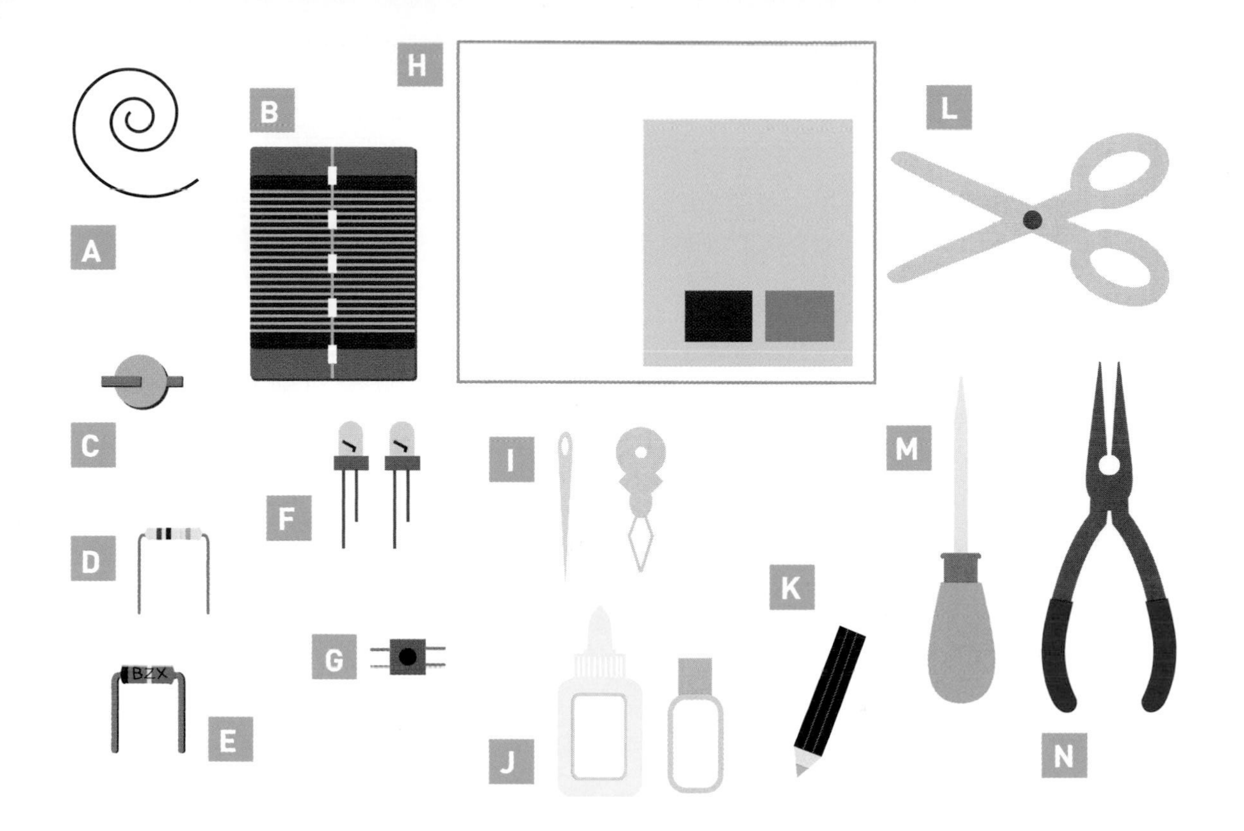

LIST OF SUPPLIES:

- A Conductive thread
- B 6 volt, 100 milliamp solar panel
- C 5 volt, 0.33 F (farad) super capacitor
- D 100 ohm resistor
- E 5239 Zener diode
- F 2 LEDs
- G Push button
- H Felt sheet
- I Sewing needle and threader
- J Fabric glue and clear nail polish
- K Chalk pen
- L Scissors
- M Awl or other tool for poking holes in felt
- L Needle-nose pliers

Make It

Now it's time to follow these steps to make your solar project:

8

1. Draw your design on a felt sheet using the chalk pen.

2. Cut out and glue together your felt pieces.

3. Mark the positions of your LEDs.

4. Using an awl, or other sharp tool, poke two holes for the legs of each LED.

5. Mark the positive and negative holes on the back of the project with the chalk pen.

6. Place the legs of the LEDs through the felt. Make sure the LEDs are facing the front and the legs are on the back.

7. Use needle-nose pliers to curl the legs on the LEDs so that you can sew to them.

8. On the back side of your project, sketch the circuit diagram with the chalk pen. Make sure to mark the positive and negative sides of the solar panel, the super capacitor, and the Zener diode.

Now you are ready to start sewing your circuit.

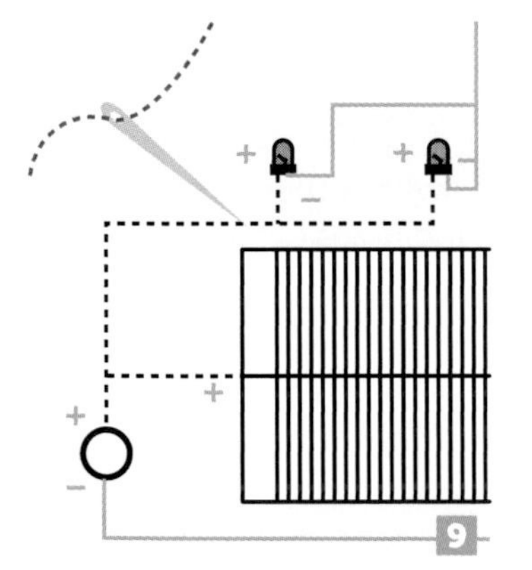

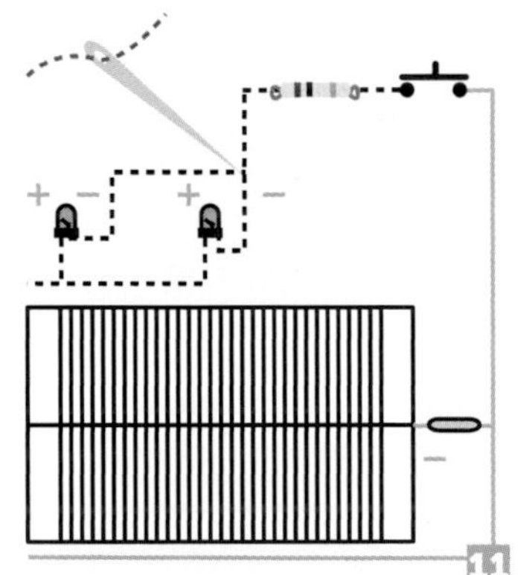

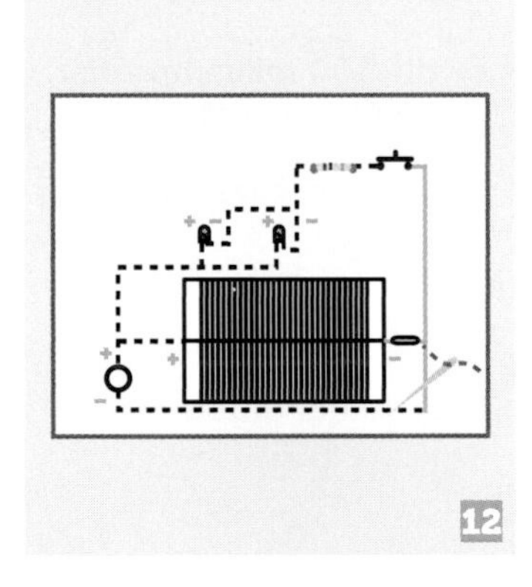

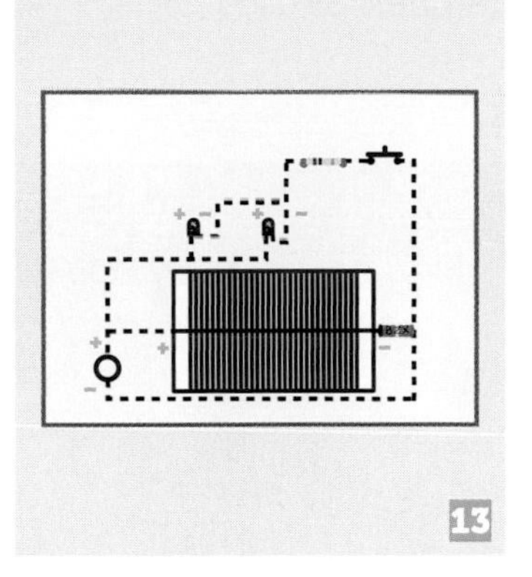

9. With conductive thread, connect the positive side of the solar panel to the positive legs of your LEDs and the positive leg of the super capacitor. Since the legs of the super capacitor are so short, you'll have to be careful as you use needle-nose pliers to bend them into hooks so that you can sew to them.

10. Use needle-nose pliers to curl the four short legs of the push button into hooks so that you can sew to them. You will end up sewing to two of the four legs; when you do, make sure to use two of the legs that are diagonal from each other on the button.

11. Using conductive thread, connect one side of the push button to the negative legs of the LEDs through a 100-ohm resistor.

12. Next, connect the negative pole of the super capacitor to the positive leg of the Zener diode.

13. Continue by connecting the Zener diode's positive leg with the free side of the push button. Then connect the Zener diode's negative leg with the negative pole of the solar cell.

14. Finish up by flipping your project over and attaching any decorations with fabric glue.

15. It takes a little bit of time to fully charge the super capacitor. Leave your project out in the sunlight for a few minutes and then press the push button to test your circuit. Your LEDs should light up until the super capacitor runs out of charge.

You'll have to put your project out in the sun to recharge the super capacitor each time that you want to impress your friends!

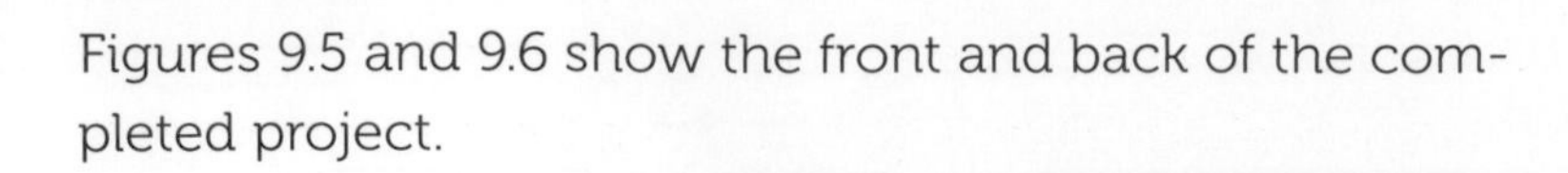

Figures 9.5 and 9.6 show the front and back of the completed project.

FIGURE 9.5: The finished Solar Sun project

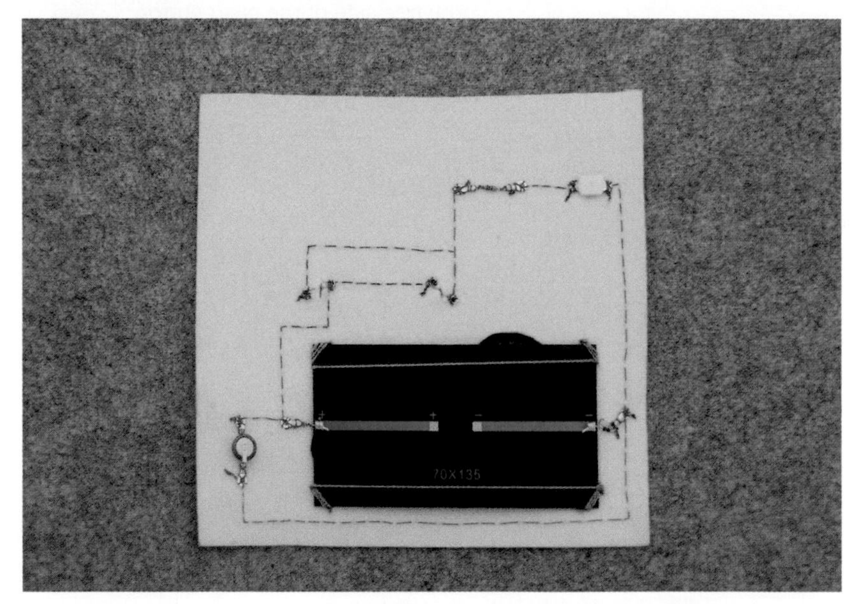

FIGURE 9.6: The back of the Solar Sun project with the circuit

Projects that use solar panels can lead to interesting discussions about sustainable energy. When you share this project, make sure to tell people how solar energy is better for the environment than batteries that you throw away.

The remainder of the figures in this chapter show other projects that use solar panels. Figures 9.7 and 9.8 show the front and back of a solar-powered house.

Solar cells are a great way to power projects without using batteries. However, projects that use solar cells have to be exposed to sunlight to work. When I was thinking about projects that would work well outdoors, I thought about making a handbag that lights up. The result is the Solar Bag project (see Figures 9.9–9.11). Round holes in the side of the bag expose solar cells that can gather electricity during the day. When it gets dark, a strip of LEDs turns on to light up

FIGURE 9.7: Solar-powered house

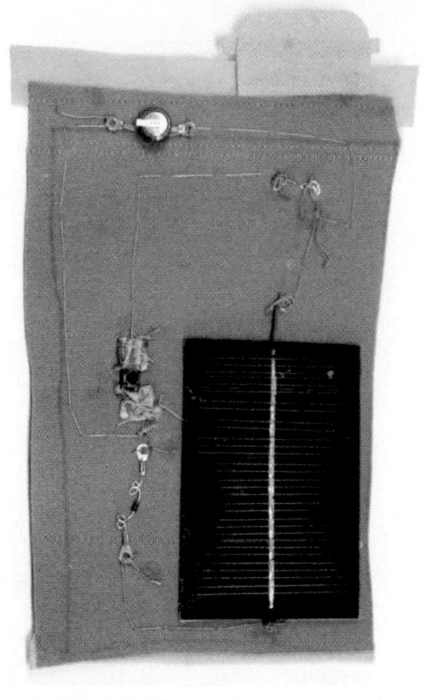

FIGURE 9.8: The solar-powered house's circuit

the night. I am still developing this project as a design experiment.

FIGURE 9.9: The lights for the Solar Bag project

FIGURE 9.10: Prototyping the circuit for the Solar Bag project

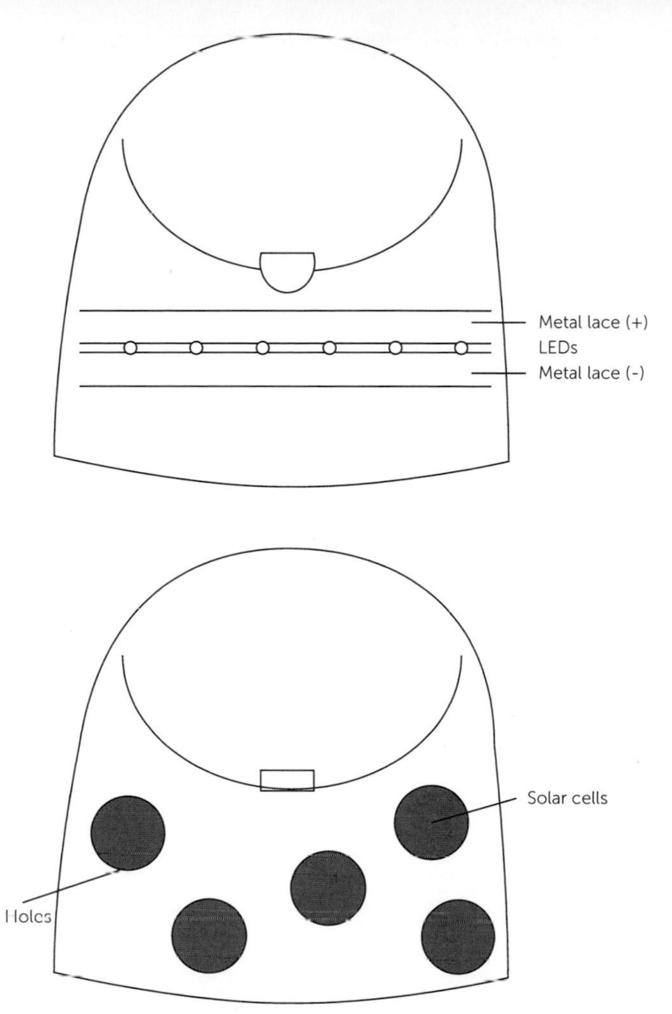

FIGURE 9.11: A sketch and plan for the Solar Bag project

What's Next?

Congratulations! Now you know how to use solar panels to make your projects more sustainable and earth friendly. For the final project, you are going to make your own board game with built-in homemade switches, connect it to a computer, and write your own computer program!

Chapter 10

Dice Board Game

PROGRAMMING WITH SCRATCH AND MAKEY MAKEY

For the final project in this book, you are going to move beyond simple circuits and into computer programming. With computers, we can make our circuits even smarter. In fact, sensors in the real world can affect what is happening on a computer screen!

In this project, you are going to make a board game that includes soft switches that you will build yourself. You will also write a computer program that responds to switches being pressed on your board game. Then you will write a computer program that rolls random digital dice and triggers some "mystery" spaces in your game.

To make the project shown in Figure 10.1, you are going to be using the Scratch programming language and the Makey Makey microcontroller, which are both really powerful and fun tools. You are also going to make some homemade fabric switches for this project. Let's learn a bit about all of these new tools and ideas before you begin.

FIGURE 10.1: Lucky Dice board game with computer-controlled dice

Scratch

If you have never programmed a computer before, Scratch is a great language to begin with. Scratch was developed at the Massachusetts Institute of Technology (MIT) Media Lab by the Lifelong Kindergarten Group. Scratch is designed to make it easy and fun for children to learn computer programming and create awesome projects that they can easily share online. Scratch makes it simple to learn how to program because the code uses blocks that you snap together instead of requiring you to write text. To make a Scratch program, you snap these blocks together much like you do when you're constructing with LEGOs. Not only is it really easy to get started, as you get better at Scratch, it is possible to make some advanced programs. You can make video games, solve difficult math problems, make animations or music videos, design and play music instruments, and much, much more. On the Scratch website, you can create your own projects for free, share those projects with the world, and even remix projects made by other Scratchers! You can get started right now and learn more at http://scratch.mit.edu.

Makey Makey

This project also uses the Makey Makey microcontroller—a small, inexpensive computer—to connect the board game to your computer and the Scratch program. The Makey Makey is a really fun device that allows you to trigger keyboard key presses by connecting alligator clips (or other conductive objects). This makes it so you can, for instance, simulate pressing the space bar by touching a piece of conductive thread and a piece of tinfoil together. Then you can write a Scratch program that reacts to these simulated key presses. To learn more about the Makey Makey and what you can do with it, check out the Makey Makey website at www.makeymakey.com/.

Fabric Switches

Your game board is going to use several homemade soft switches to trigger key presses, like the space bar and the direction keys (up, down, left, and right), through your Makey Makey. If you press the middle of these fabric switches, electricity is able to flow from one piece of conductive material to the conductive material on the other side of the switch. After gathering the following materials, you can follow the steps listed here to make a simple fabric switch for your own projects.

MATERIALS YOU WILL NEED:

Conductive fabric or tinfoil

Felt

Thick fabric, felt, or a thin sponge

1. Glue two strips of conductive fabric or tinfoil to two squares of felt so they appear as they do on the left side of Figure 10.2. Make sure to leave some of the conductive material hanging off to the side so that you have two places to connect your circuit to the switch.

2. Cut a hole in the middle of a square of thick felt (see the middle image in Figure 10.2). You can also use two layers of regular felt or a thick sponge. The thickness of this material has to be enough to keep the two conductive fabric pieces from touching each other in the next step.

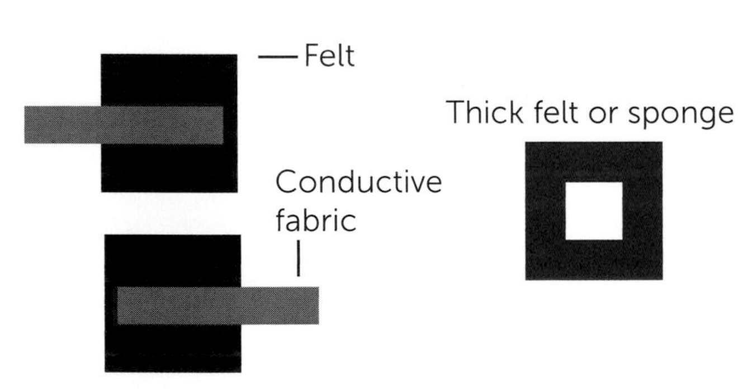

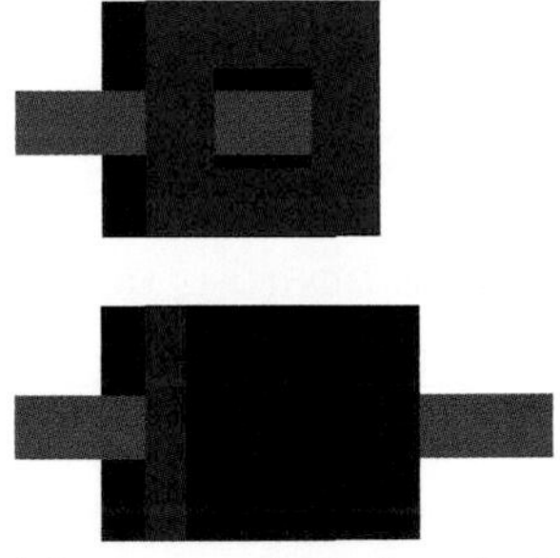

FIGURE 10.2: How to make a fabric switch

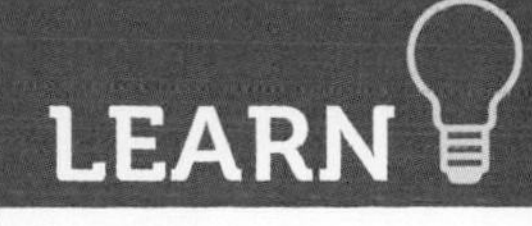

3. Glue the piece from Step 2 in between the two pieces from Step 1 (see the top-right image in Figure 10.2). Start by gluing the thick felt or sponge piece on top of one of pieces of the conductive material (see Figures 10.3 and 10.4). Then glue the second piece from Step 1 on top of the thick felt or sponge piece from Step 2. The finished switch is shown in the bottom-right image in Figure 10.2. The two strips of conductive fabric should face each other but not touch each other unless you press down on the switch.

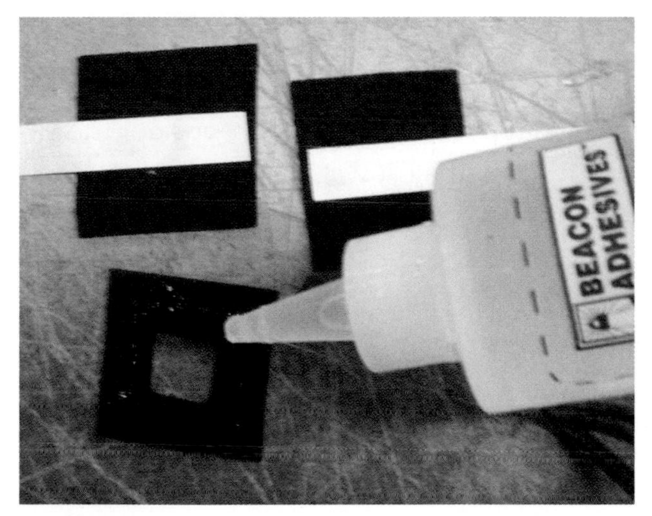

FIGURE 10.3: Apply glue to the thick felt or sponge that goes in between the two pieces of conductive material.

FIGURE 10.4: This is the switch before gluing on the top layer.

For this chapter's project, you are going to build similar switches into your board game.

Use Your Imagination

What kind of board game do you want to make? I was inspired by the classic game *Chutes and Ladders*. In my game, players "roll" the digital dice in a computer program by pressing a fabric switch on the bottom-right corner of the board game. If a player lands on a space that contains a question mark, they can press a fabric switch on that space and get a mystery mission or quiz question.

Now that you understand how my game works, think of your own version of a board game that can use fabric switches and a computer program.

Before you start sketching your board game, consider the size of the board. If it's too small, it will be hard to fit playing pieces on the spaces. If you can't find a piece of felt that is large enough for your game, you can use a piece of cardboard instead. The game that I made is about two feet tall and one foot wide. For my sketch, I taped three pieces of paper together to make a long rectangle (see Figure 10.5), but yours does not have to be that big.

Plan It Out

On your sketch, to help plan your circuit, mark off the positions where the fabric switches will go. I used one switch to trigger the digital dice to roll in my Scratch computer program. I also positioned four "mystery" switches on some of the spaces in the game. The top sides of these switches use conductive thread to connect to key triggers on the Makey Makey. I used the up, down, left, and right arrows and the space bar for these trigger keys. In my setup, a tinfoil sheet

FIGURE 10.5: A sketch of the SuperGiantDice board game

underneath the game board acts as the bottom side of each switch. That sheet is connected to the "earth" strip on the Makey Makey. When any of the switches are depressed, the conductive thread touches the tinfoil underneath the board and closes the switch between the key trigger and the "earth" on the Makey Makey (see Figure 10.6).

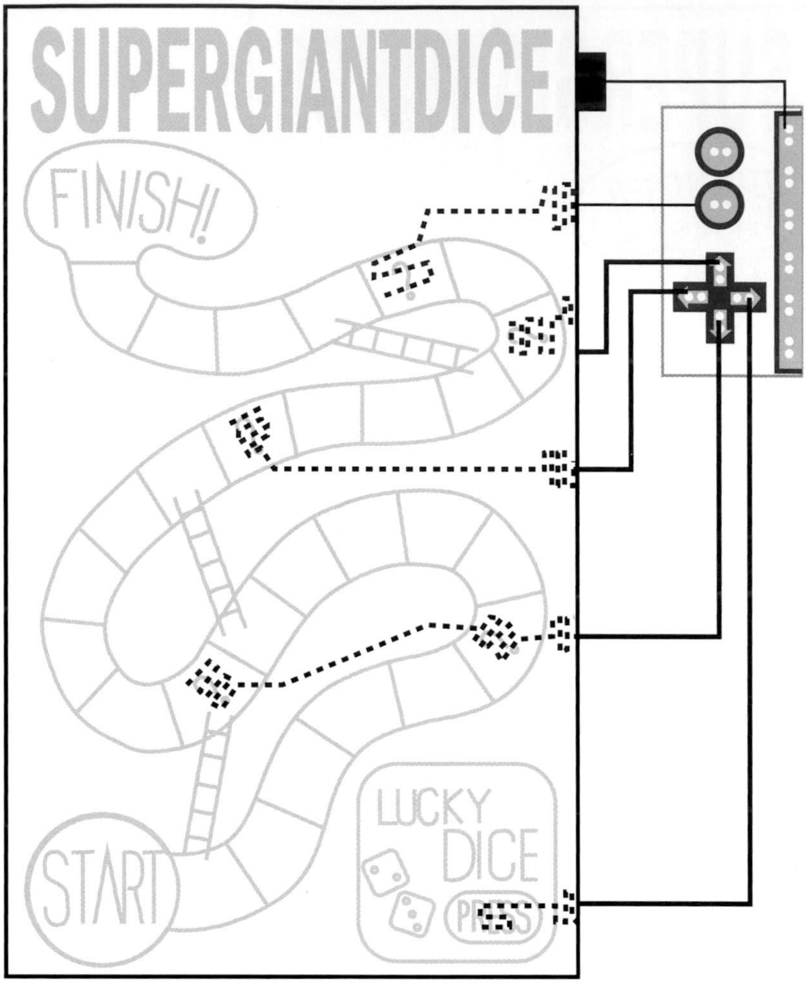

FIGURE 10.6: The board game circuit sketch with the switches connected to the Makey Makey

Preparation

Once you have your sketch and plan for the game all ready, you can gather your supplies, make your game board, connect the circuit, and write your computer program.

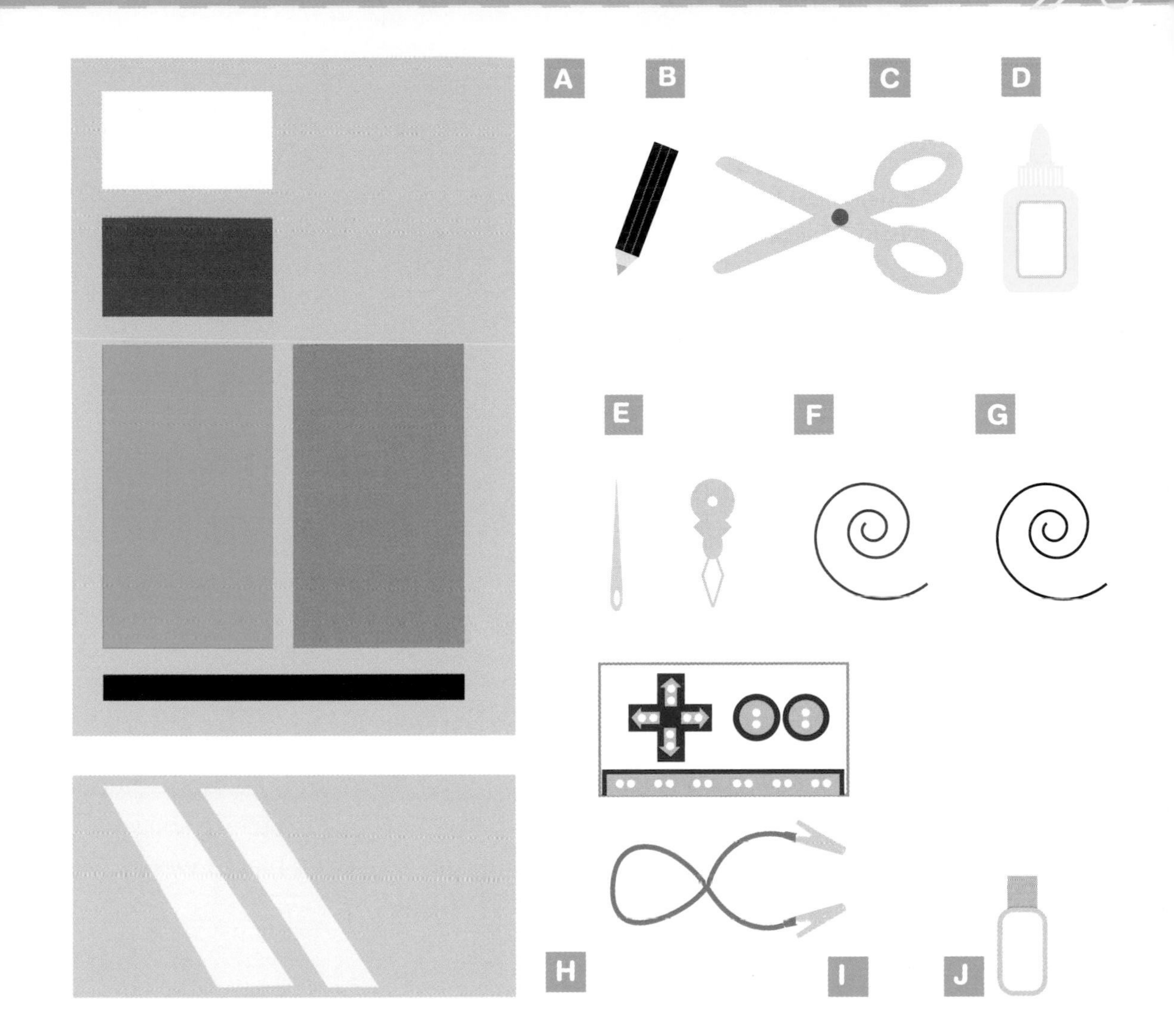

LIST OF SUPPLIES:

- A Felt sheet (If you can't find a large enough sheet, you can use cardboard for the board)
- B Chalk pen
- C Scissors
- D Glue
- E Needle and threader
- F Thread
- G Conductive thread
- H Aluminum foil or conductive fabric
- I Makey Makey with alligator clips
- J Nail polish

Make It

Follow these steps to make a board game that connects to a computer:

1. Based on your paper drawing, use a chalk pen on a large sheet of felt (or cardboard) to sketch your project.

2. Cut out any felt pieces for your game and use fabric glue to connect them to your board. You can also use embroidery stitches for words or decorations. Leave the places where your switches are going to go blank for now.

3. Carefully cut holes on your game board under the places where you want to put your switches.

4. Using conductive thread, sew back and forth to create a conductive surface that will act as the top part of your switch.

5. Then lead that thread to the edge of your game board so that you can connect an alligator clip to the end. Be careful that the thread does not go through the game board to the other side unless you press the switch down.

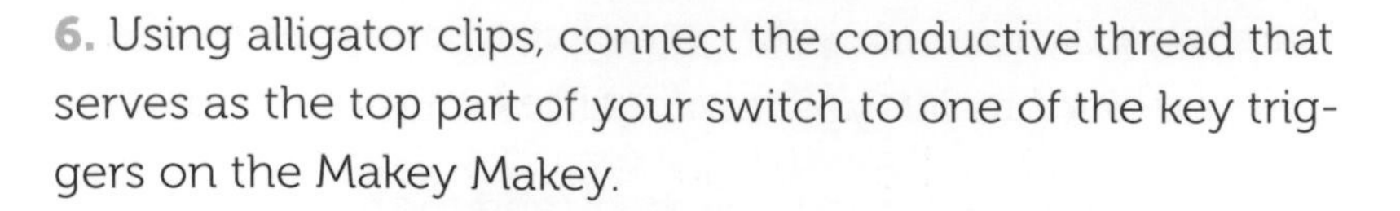

6. Using alligator clips, connect the conductive thread that serves as the top part of your switch to one of the key triggers on the Makey Makey.

7. Repeat steps 4–6 for each of your switches.

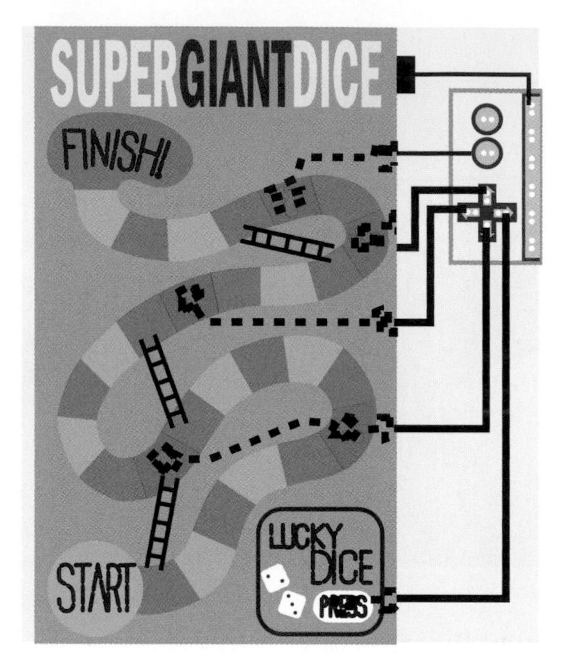

8. Place a sheet of aluminum foil or conductive fabric underneath your game board.

9. Using an alligator clip, connect the aluminum foil or conductive material underneath your game board to the "earth" metal strip on the bottom of your Makey Makey. Take a look at the completed game.

10. Now it's time to test your switches by connecting your Makey Makey to a computer with the included USB cable. If any windows pop up when you plug in the Makey Makey, you can just close them.

11. Open any text or writing program. You should be able to trigger the space key and direction arrows by pressing your switches.

12. Once you know that you switches are working, it is time to write your Scratch program. Go to http://scratch.mit.edu, create an account if you don't have one, and start a new project.

13. You can delete the Scratch Cat sprite by choosing the scissors tool at the top of the screen and then clicking on the cat.

14. Begin programming by creating a new sprite for your dice. Make six costumes for this sprite, one for each possible dice roll.

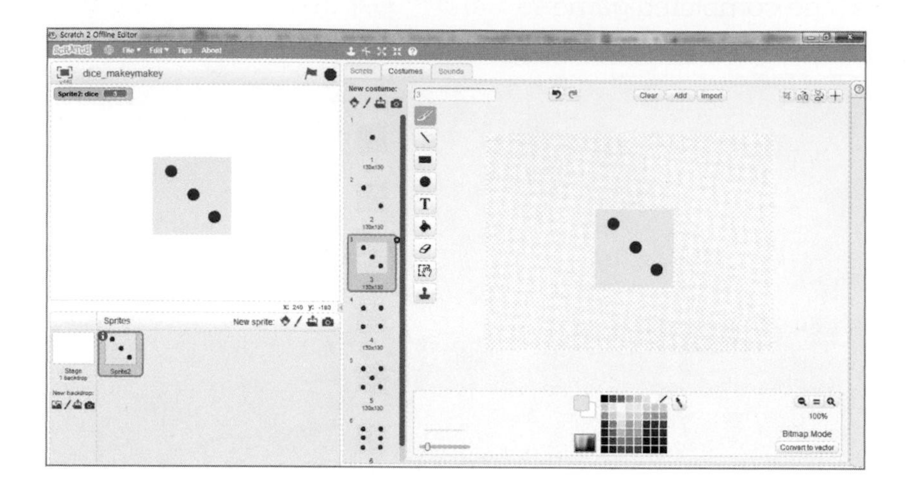

15. In the Data scripts, create a new variable named **dice**.

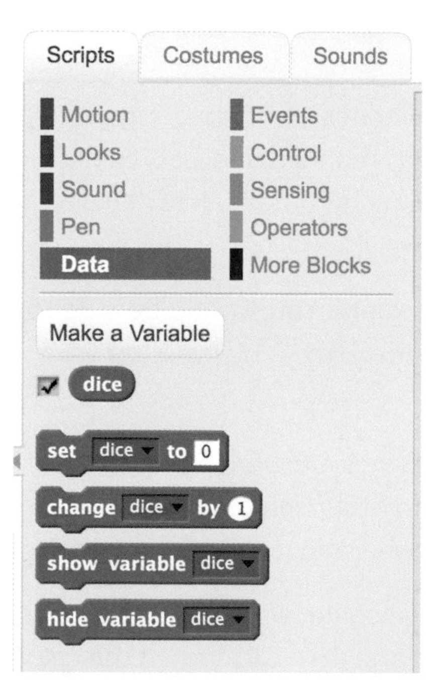

16. Now you will create a script for your dice sprite that says, "When the space key is pressed, switch the costume randomly and play a rattle sound 10 times. After that, choose a random number between 1 and 6, and then display that costume for the dice."

17. Create some Scripts for your "mystery" switches. These switches are connected to the up, down, left, and right key triggers on the Makey Makey, so you should create scripts that are triggered by those keys. Be creative! Here are some examples.

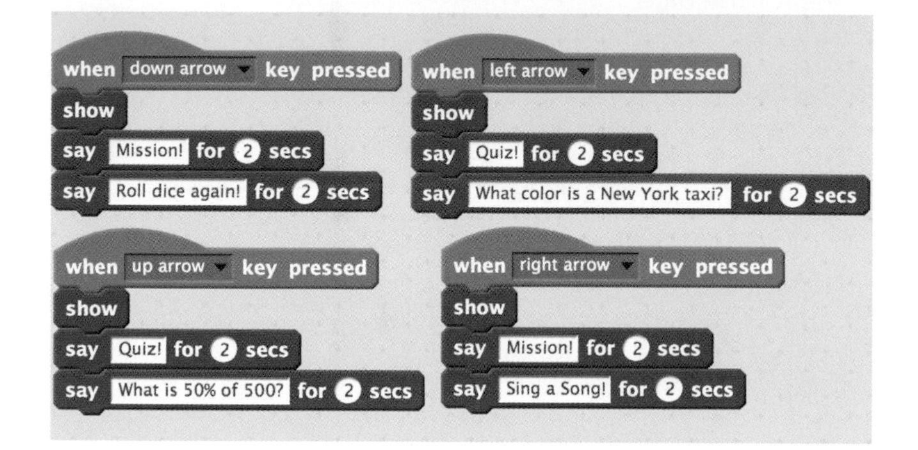

18. Make sure that your Makey Makey is connected and then test that your switches trigger the scripts by pressing each one.

Figures 10.7 and 10.8 show different views of the completed game.

FIGURE 10.7: The board game with switches connected to the Makey Makey

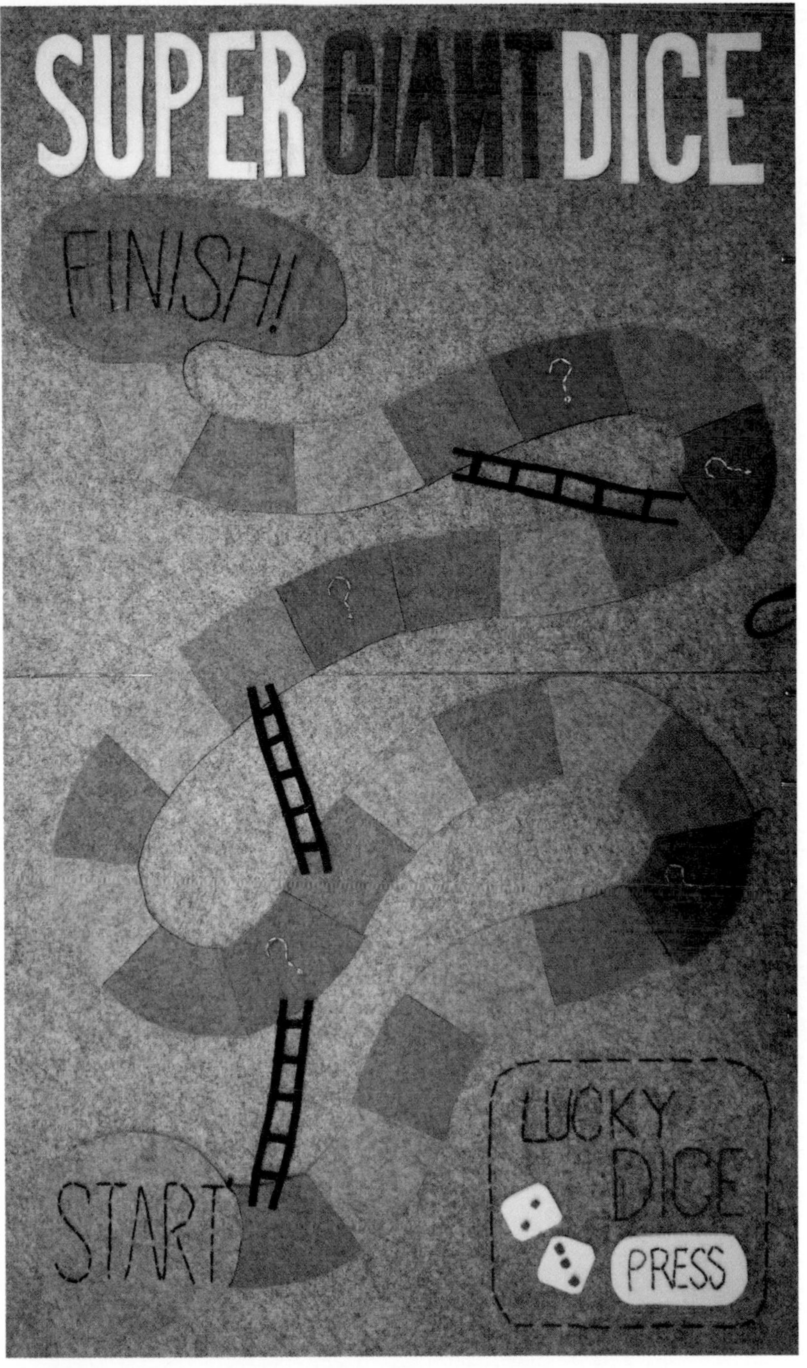

FIGURE 10.8: The finished board game

Invite some friends over to play-test your new game! Make sure to share your Scratch program on the Scratch website. The Scratch community is a fantastic place in which to share projects with other enthusiastic game designers. You can find the Scratch program associated with the project in this chapter at https://scratch.mit.edu/projects/101845048.

You can make your game even more fun by adding sounds or animations to your Scratch program. To learn more about other things that you can do with Scratch, check out some of the resources on the Scratch Help page at https://scratch.mit.edu/help/.

Designing games this way is endless fun. Figure 10.9 shows my daughter working on a game.

Josh Burker, a middle school technology teacher, had yet another game idea. He made this version of the classic game *Operation!* (see Figure 10.10) using Scratch and the Makey Makey. You can find a guide for how to make this game

FIGURE 10.9: My daughter designing her own board game

on the Makey Makey website at http://makeymakey.com/guides/pdfs/MaKeyMaKeyScratchOperationGame.pdf.

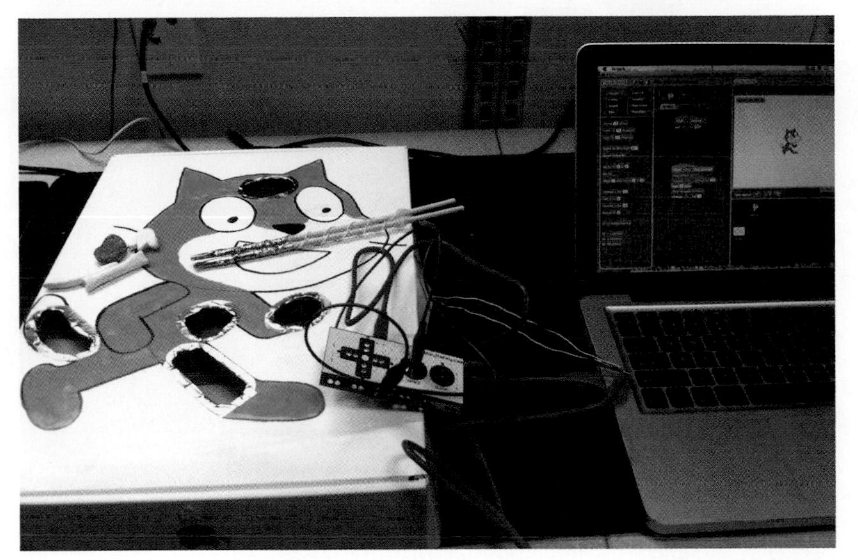

FIGURE 10.10: Josh Burker's *Operation!* remake

What's Next?

Congratulations! You've now finished all the projects in this book! Just think back on what you've accomplished—you started with a simple LED circuit and worked all the way up to writing your own computer program! By now, you know that the only way to get better at sewing and working with electronics is to continue practicing. I challenge you to keep making circuits as you imagine new ways to use the knowledge and skills that you gained by working through the projects in this book.

Would you like one more challenge before you go off on your own? Take a look at the appendix to see how you can make the moonlit flowers we discussed in Chapter 3. Good luck with all your switch-making endeavors!

Appendix A

The Moonlit Flower

USING SNAPS AS A SWITCH

At the end of Chapter 3, I described the moonlit flower, a wearable light-up ring that uses a metal snap button as an on/off switch. This project was a bit advanced to put at the beginning of the book, so I decided to share the instructions here in an appendix. If you completed several of the projects in the book and you feel comfortable sewing and making circuits with conductive thread, you'll enjoy this project. Here is how you can make a moonlit flower bracelet that lights up.

Preparation

You need to gather the following materials before you begin the process of making the moonlit flower bracelet.

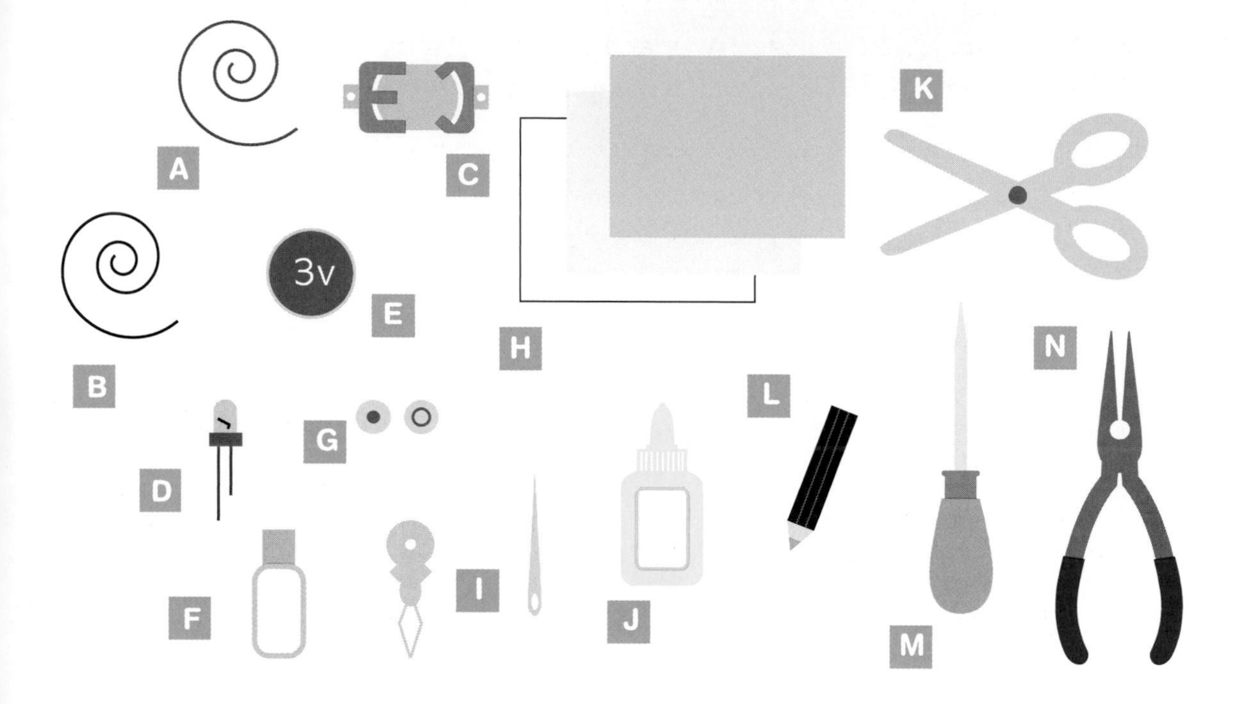

LIST OF SUPPLIES:

A Regular, non-conductive, thread

B Conductive thread

C Battery holder

D LED

E 3V coin cell battery

F Clear nail polish

G Metal snaps

H Felt sheets

I Sewing needle and needle threader

J Fabric glue

K Scissors

L Chalk pen

M Awl

N Needle-nose pliers

Make It

Since this is a project that you can wear, take your time and make sure to plan out the length of the strap and the placement of the flower, battery holder, and the snaps before you begin. It is a great idea to make a paper and tape prototype first so that you can confirm that you are happy with the size.

Pay extra attention when you are sewing on the snaps. It is easy to put them on upside down or on the wrong side of the strap. Double-check that they will snap together correctly before you sew them down. Follow these steps to create your flower bracelet:

1. Begin the process of making your moonlit flower bracelet by first cutting out two versions of the five-petal flower shape. If you make the flowers too small, it will be hard to stitch them neatly. We recommend that the flowers be about two inches wide.

2. Now measure and cut out your arm strap. The length of your strap will depend on the thickness of your wrist. Measure the circumference of your wrist with a tape measure or with a piece of string and a ruler. Make sure you add one inch to your measurement to account for the overlap and the snap.

Make the strap about one inch wide with a rounded end.

3. Fold the felt across the center of a flower petal and then, using regular (non-conductive) thread, sew just above the folded crease on the back of the petal. Make sure to stop sewing where the petal meets the center of the flower. Do not sew across the center. Repeat this for all the petals. Sewing these creases on the bottom of the petals helps the petals pop out into three dimensions.

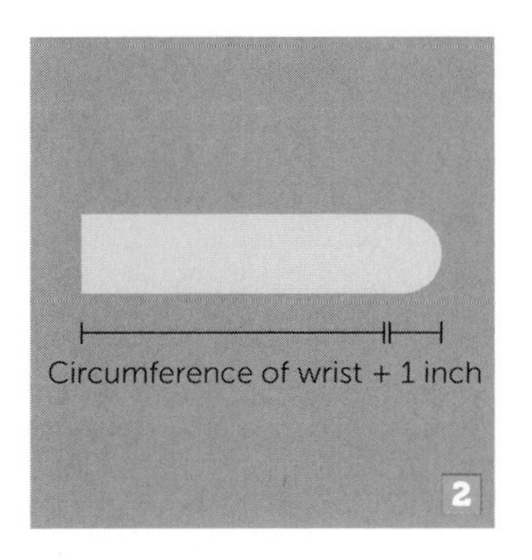

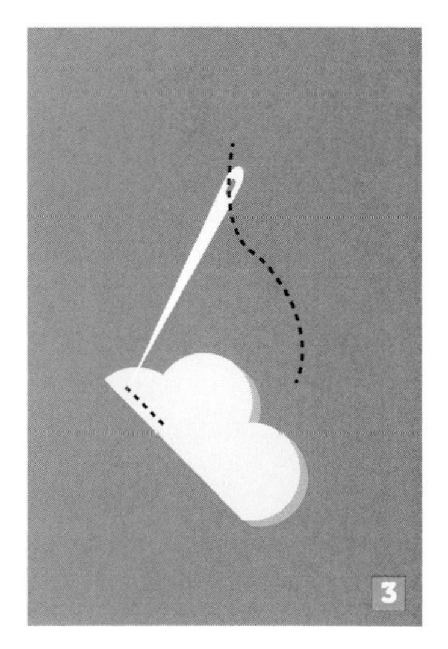

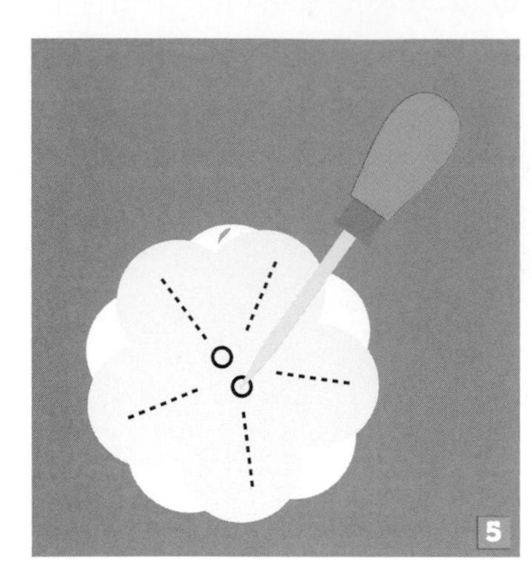

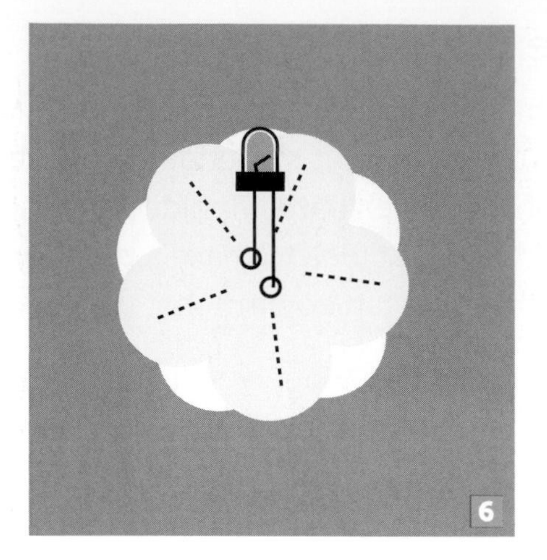

4. After you're done sewing all the petals on both flower shapes, sew the two flower shapes together with regular, non-conductive thread. Offset the petals slightly so that they are not stacked directly on top of each other. Then loop the thread through both flowers a few times near the middle to make your project look like a flower. Tie off the thread at the bottom of the flower.

5. Now that you've got your flower prepared, it's time to make it light up! Start by using an awl or another sharp poking tool to carefully make two holes for the LED legs in the center of the flower.

6. Poke the LED legs through the holes with the LED facing up in the center of the flower and with the legs poking out the back of the flower. Use a chalk pen to mark which hole has the positive or longer leg of the LED coming out.

7. Use needle-nose pliers to curl the legs of the LEDs so that you can sew to them. Make sure you remember which one is the positive leg and which one is the negative leg.

8. Now it's time to glue your battery holder to the strap. First check the placement of your battery holder. *If you point the rounded end of the strap*

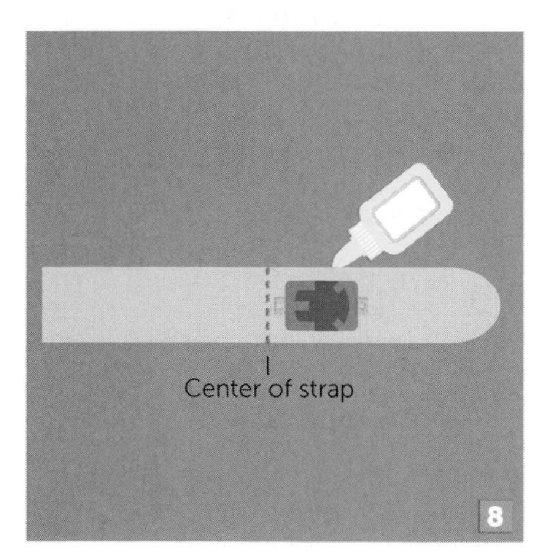

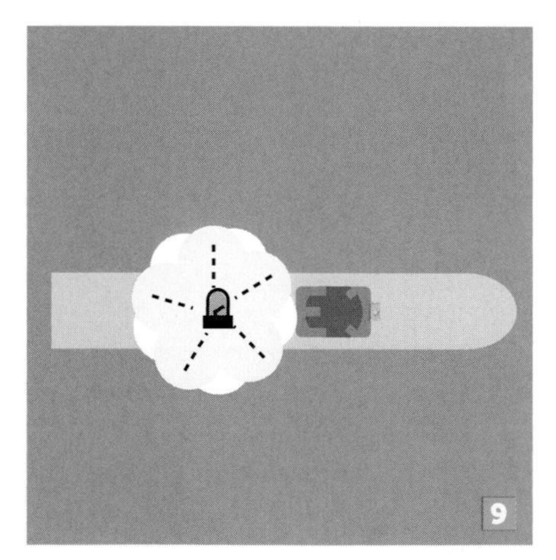

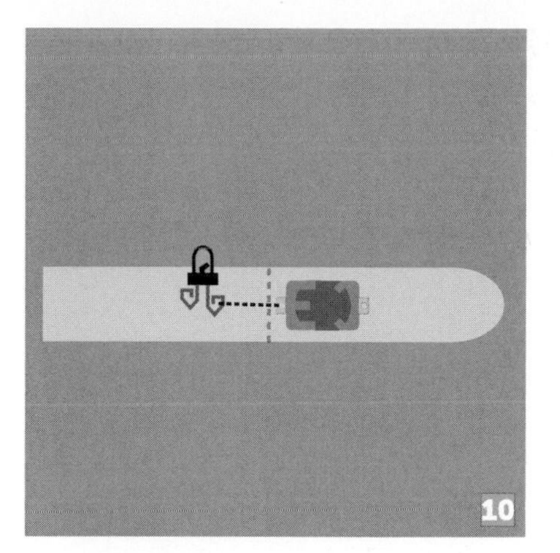

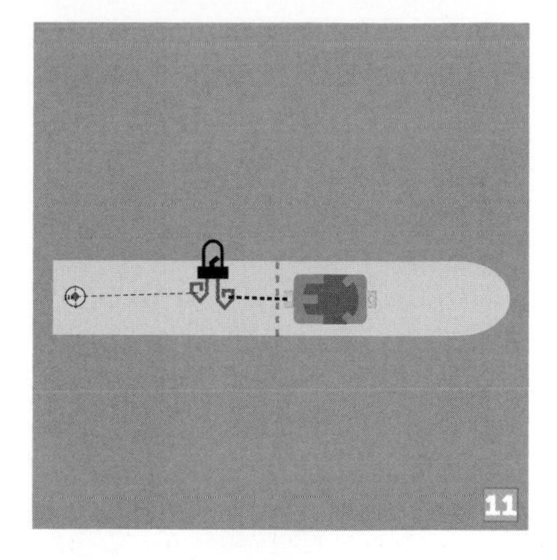

toward the right, the battery holder should be just to the right of the center line of the strap. Then paste the battery holder in position on the felt with glue.

9. Now it's time to attach your flower to your wrist strap. Place the flower with the LED just to the left of the battery holder.

10. Using conductive thread, connect the positive leg of the LED to the positive side of the battery holder. *Note that in this illustration, the flower has been removed for clarity.*

11. Using conductive thread, connect one snap to the outside of the rectangle end of the strap. Make sure the snap is facing the right way. It is too easy to sew snaps on upside down! Poke the needle through the inside of the bracelet and then push the needle through a hole on the snap. Loop two or three times under and around the hole on the snap so that you get a good electrical connection. Then continue to sew down the snap by looping the needle and thread through the other three holes. Once the snap is secure, continue sewing with the same piece of thread and loop it around the negative leg of the LED a few times before tying it off. Use a bit of clear nail polish on the knots so that they do not come unraveled. *Note that again, the flower has been removed for clarity.*

12. Finally, using conductive thread, connect the other half of the snap to the rounded end of the strap. Position the snap so that it goes on the inside of the strap. The rounded end of the strap will overlap the rectangle end and snap with its mate. *Note that this snap is shown on the outside of the strap for clarity.* Double-check that you are sewing it to the right place and that it will snap shut before you sew it down. Use the same technique as in step 11 to sew down the snap and then continue on with the same piece of conductive thread and connect it to the negative side of the battery holder. Dab a bit of clear nail polish on the knots so that they donot come unraveled.

13. Now put your 3V coin battery into the battery holder.

14. Fasten the snap to close your switch; the light in the flower should now turn on!

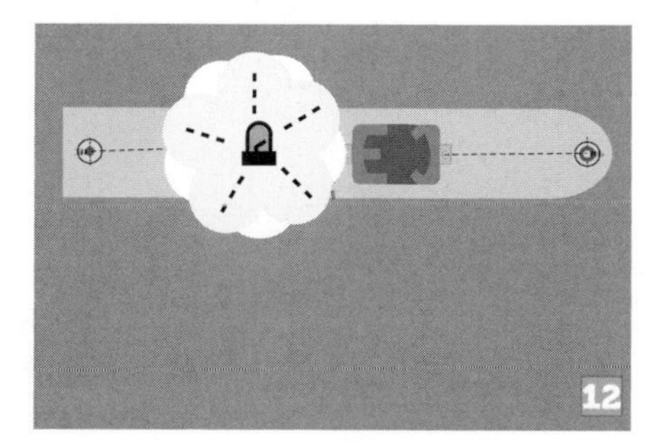

This project described here is a bracelet, but you could adapt this to add moonlit flowers to a belt, a shirt, or a bag. You can attach moonlit flowers to anything you want! Also, feel free to experiment with different materials for your flowers. Figure A.1 shows two mesh fabric flowers, one on a pin and one on a bracelet.

FIGURE A.1: A mesh moonlit flower bracelet

Index

D

E

N

O

P

T

U

V

W

Z